Air Quality in America

Air Quality in America

A Dose of Reality on Air Pollution
Levels, Trends, and Health Risks

Joel M. Schwartz and
Steven F. Hayward

The AEI Press

Publisher for the American Enterprise Institute

WASHINGTON, D.C.

Distributed to the Trade by National Book Network, 15200 NBN Way, Blue Ridge Summit, PA 17214. To order call toll free 1-800-462-6420 or 1-717-794-3800. For all other inquiries please contact the AEI Press, 1150 Seventeenth Street, N.W., Washington, D.C. 20036 or call 1-800-862-5801.

NRI NATIONAL
RESEARCH
INITIATIVE

This publication is a project of the National Research Initiative, a program of the American Enterprise Institute that is designed to support, publish, and disseminate research by university-based scholars and other independent researchers who are engaged in the exploration of important public policy issues.

Library of Congress Cataloging-in-Publication Data

Schwartz, Joel, 1965-
  Air quality in America : a dose of reality on air pollution levels,
trends, and health risks / by Joel M. Schwartz and Steven F. Hayward.
     p. cm.
  Includes bibliographical references and index.
  ISBN-13: 978-0-8447-7187-8
  ISBN-10: 0-8447-7187-2
  1. Air quality—United States. 2. Air—Pollution—United States.
3. Air quality management—Government policy—United States. I. Hayward,
Steven F. II. Title.

    TD883.S33 2007
    363.739'220973—dc22

                        2007021929

11 10 09 08 07   1 2 3 4 5

Printed in the United States of America

Contents

List of Illustrations

Acknowledgments

Steve Hayward first suggested this project to me a few years ago—a distinction that has earned him both my gratitude and wrath, but mostly my gratitude. I've tried to return the favor by roping him in as a coauthor.

I've benefited from discussions with many experts and scholars in environmental science, health, economics, law, policy, and politics. I'd particularly like to thank Jonathan Adler, Gary Bishop, Buzz Breedlove, Roy Cordato, Chris DeMuth, Ted Gayer, Indur Goklany, Ken Green, Bob Hahn, Matt Kahn, Doug Lawson, Marlo Lewis, Fred Lipfert, Randall Lutter, Suresh Moolgavkar, Andy Morriss, Lynn Scarlett, David Schoenbrod, Randy Simmons, Brett Singer, Don Stedman, and Rick Stroup.

Buzz Breedlove, Tom Darlington, Chris DeMuth, Ken Green, Henry Olsen, and Don Stedman read and provided helpful comments on all or portions of various drafts of the manuscript.

Dennis Kahlbaum of Air Improvement Resource (AIR) performed quality control and analysis of much of the ambient air pollution monitoring data used to estimate national trends in pollution levels. Dennis also performed the analysis of how many people live in areas where air pollution levels violate federal standards. This project would not have been possible without his extraordinary expertise and attention to detail in compiling, quality checking, and analyzing large data sets. I'm also grateful to Dennis as well as Tom Darlington and Jon Heuss of AIR for their encouragement and advice.

I thank the publications staff at AEI, including Tim Lehmann, Jennifer Moretta, Kathy Swain, and Lisa Ferraro Parmelee, for their careful copyediting and typesetting of the manuscript.

Needless to say, any remaining errors of fact or logic are of course my own.

I wrote this book as a visiting scholar at the American Enterprise Institute. I'm grateful to the Searle Freedom Trust and AEI's National Research Initiative for financial support and to Kim Dennis, Henry Olsen, and Chris DeMuth for their continued support, counsel, and encouragement. Ryan Stowers was also a constant source of encouragement during his tenure at the National Research Initiative.

Most of all I thank my wife, Laura Mahoney, and my children, Benjamin and Helen, for their patience, support, and devotion over the years in which this project came to fruition.

Introduction

Americans are sensitive about air pollution, and no wonder—the air we breathe is perhaps the most elemental aspect of environmental quality. Opinion polls routinely find that a majority of Americans believe air quality has deteriorated and will worsen in the future, and that most people face serious risks from air pollution.[1]

Public and elite perception of air pollution levels, trends, and health risks, however, is virtually the opposite of reality. America reduced air pollution dramatically throughout the twentieth century to only a fraction of past levels, and the country now enjoys relatively good air quality. Even the worst areas have far better air quality than was typical of American cities during the 1950s, '60s, or '70s.

A wide array of data attests to our success in getting rid of most air pollution and to the continuing decline in emissions from motor vehicles and industry. Already-adopted measures ensure the elimination of most remaining emissions during the next two decades. Furthermore, Americans were improving air quality in their communities for decades before the federal government nationalized air regulation with the Clean Air Act Amendments of 1970. These improvements resulted from a combination of market forces that encouraged technological advancement and increased energy efficiency, common-law nuisance suits, and local and state regulation.

Several factors account for public and media misperceptions about air quality. Air pollution comes in many forms, from many sources, and varies over time and location, even within a given metropolitan area. Regulatory standards and requirements are complex and arcane. Progressive tightening of standards for some pollutants can make pollution seem to be increasing even when it is declining. These intricacies can make the topic a difficult one for the layperson to follow.

However, the most important factor in public misperception is the role of environmental groups and regulatory agencies, which exaggerate air pollution levels and health risks and often obscure positive trends, and news media that report these misleading representations with little or no critical review. Even health scientists often misrepresent the results and weight of the evidence from air pollution health studies, creating an appearance of much greater risk than the research actually suggests.

As a result, public fears are out of all proportion to the, at worst, minor health risks posed by current, historically low air pollution levels, and there is widespread but unwarranted pessimism about the nation's prospects for further improvement.

Once might ask why it matters if Americans' pessimism and fears about air pollution are groundless. After all, regardless of the size of the risks, doesn't every reduction in pollution make people better off?

This would be true if air pollution were the only risk we faced, and if reducing it were free. But the question isn't whether we would prefer less air pollution. All else equal, of course we would. In the real world, we can't keep all else equal. We face many threats to our health and safety and have limited resources with which to address them. If we devote excessive resources to one exaggerated risk, we are less able to counter other, genuinely more serious risks, or to spend our resources on other important needs and desires, such as health care, education, vacations, and housing. Highlighting small risks diverts public attention from potentially more serious problems, and unwarranted alarmism causes unnecessary fear.

Indifference to public misperceptions about air pollution would also be reasonable if the Clean Air Act (CAA), which created our current system of federal control of air pollution policy, were a resounding success. Such a view is mistaken. Air quality has indeed improved since the 1970 passage of the CAA. But it was improving at about the same pace for decades before the act was passed, and without the unnecessary collateral damage caused by our modern regulatory system.

While air quality has greatly improved over the last few decades, we've paid far more than necessary to get there. A few emission-reduction requirements—mainly for motor vehicles and power plants—account for the vast majority of improvements since passage of the Clean Air Act. Yet most regulatory activity is unrelated actually to reducing emissions and instead involves creating and complying with administrative and other process require-ments. Furthermore, our regulatory system often devotes great resources toward small, expensive, slow, and ineffective pollution-reduction measures, while ignoring opportunities for large, cheap, and rapid improvements. And as the potential health and other benefits of each increment of pollution reduction have become ever smaller, the incremental costs have continued to grow.

The Five Main Findings of This Study

We wrote this book to demystify air pollution levels, trends, prospects, and health risks through a wide-ranging analysis of long-term national pollution monitoring data and a review of atmospheric and health studies. We assess the actual number of people living in areas that violate federal pollution standards, and the health risks posed by recent pollution levels. We also discuss whether our regulatory system is an especially good way to enhance air quality, and how it might be improved. Looking toward the future, we explain why we can expect air quality to continue to improve. The nation's air quality is far better and the health risks of current pollution levels far lower than is commonly believed. Five main findings support this overall conclusion:

The nation has sharply reduced air pollution levels, despite large increases in nominally "polluting" activities. Figure I-1 on the follow-ing page compares trends in air pollutants with trends in motor-vehicle transportation, energy production, and economic activity from 1980 to 2005. Note that despite large increases in driving, energy production, and economic activity in general, the nation nevertheless achieved large declines in air pollution levels.

These reductions in ambient air pollution levels resulted in large increases in compliance with Clean Air Act air quality standards. Virtually the entire nation (more than 99 percent of monitoring locations) now meets federal air quality standards for carbon monoxide, sulfur dioxide, nitrogen dioxide, and lead. Levels of all of these pollutants continue to decline. Ozone and fine particulates (PM$_{2.5}$; see "A Guide to Different Types of Air Pollution" on page 6) also continue to drop, but exceedance of standards

FIGURE I-1

TRENDS IN MILES DRIVEN, ENERGY PRODUCTION, AND ECONOMIC
ACTIVITY VS. TRENDS IN AIR POLLUTION LEVELS, 1980–2005

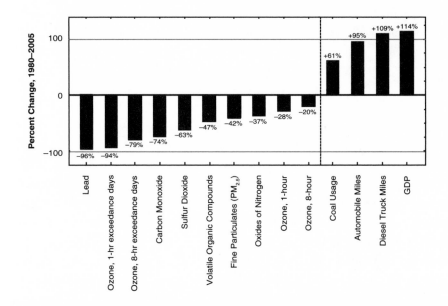

SOURCES: Bureau of Economic Analysis, *National Economic Accounts*, U.S. Department of Commerce, http://www.bea.doc.gov/bea/dn/nipaweb/SelectTable.asp (accessed November 15, 2006); Bureau of Transportation Statistics, *National Transportation Statistics*, http://www.bts.gov/publications/national_transportation_statistics/ (accessed November 15, 2006); D. O. Hinton, J. M. Sune, J. C. Suggs, et al., *Inhalable Particulate Network Report: Operation and Data Summary* (Mass Concentrations Only, Volume I, April 1979–December 1982 (Research Triangle Park, N.C.: Environmental Protection Agency, November 1984); D. O. Hinton, J. M. Sune, J. C. Suggs, et al., *Inhalable Particulate Network Report: Data Summary* (Mass Concentrations Only), Volume III, January 1983–December 1984 (Research Triangle Park, N.C.: Environmental Protection Agency, April 1986); Energy Information Administration, *Annual Energy Review* 2005, July 2006, http://www.eia.doe.gov/emeu/aer/pdf/aer.pdf (accessed November 15, 2006); Environmental Protection Agency, *Air Emission Trends*, http://www.epa.gov/airtrends/econ-emissions.html (accessed November 15, 2006); Environmental Protection Agency, *AirTrends*, http://www.epa.gov/airtrends/ (accessed November 15, 2006); Environmental Protection Agency, *Airdata: Reports and Maps*, http://www.epa.gov/air/data/reports.html (accessed November 15, 2006).

NOTES: The value for volatile organic compounds (VOCs) represents the change in estimated emissions, since no long-term measurements of ambient levels of total VOCs are available. The trends for lead, sulfur dioxide, nitrogen dioxide, and particulates are based on annual average levels. The trends for carbon monoxide and ozone are based on peak daily levels (second highest annual value for 1-hour ozone and carbon monoxide, and fourth highest annual value for 8-hour ozone). The graph also provides the percentage change in the average number of days per year that U.S. ozone monitoring sites exceeded the 1-hour and 8-hour ozone standards. For PM2.5, the base-year data were collected at various times during 1979–83, as part of an EPA special study, rather than only during 1980.

for these pollutants is more widespread. As of the end of 2006, 15 percent of the nation's monitoring sites violate the 8-hour ozone standard and 14 percent violate $PM_{2.5}$ standards.[2] Still, this is far better than 1980, when about 75 percent of monitors violated the 8-hour ozone standard, and 90 percent violated $PM_{2.5}$ standards. These ozone and PM violation rates pertain to the tougher standards adopted by the Environmental Protection Agency (EPA) in 1997.[3] Only about a third as many monitoring locations violate EPA's earlier ozone and PM standards (the 1-hour ozone and the PM_{10} standards).

These air quality improvements represent a long-term trend that predates the 1970 Clean Air Act by several decades. We explore long-term pollution trends in detail in chapters 1 through 4, partly as a reference and partly to offer perspective on the deeper sources of long-term air quality improvement. These included technological progress, economic growth, common-law nuisance suits, and, increasingly as the twentieth century progressed, local and state regulation. Federal regulation has been responsible for most air quality improvement since 1970 not because it was necessary, but because the federal government seized control of air pollution policy from states, cities, common-law courts, and market actors who were addressing air pollution for decades before the federal government got involved.

Areas of the nation with the highest pollution levels have improved the most. Pittsburgh was the nation's smokiest city in the early 1900s. But Pittsburgh had eliminated at least three-quarters of its particulate pollution by 1970, when the federal government took over control of air quality policy. Since 1970, a few areas of California, mainly parts of the Los Angeles, Bakersfield, and Fresno metropolitan areas, have had by far the highest ozone and $PM_{2.5}$ levels in the country. But most of this air pollution has been eliminated.

The worst areas of Los Angeles exceeded the 1-hour ozone standard more than 150 days per year in the 1970s and early 1980s. Today the maximum exceedance rate is now under twenty-five days per year, and peak ozone levels have declined more than 60 percent. Los Angeles, Bakersfield, and Fresno exceeded the annual $PM_{2.5}$ standard by as much as a factor of three in the early 1980s. Today, these areas are only about 30 to 40 percent above the standard.

A Guide to Different Types of Air Pollution

EPA regulates six principal pollutants under the Clean Air Act, known as "criteria pollutants." They are carbon monoxide (CO), lead (Pb), sulfur dioxide (SO_2), nitrogen dioxide (NO_2), ozone (O_3), and particulate matter (PM).[4] The EPA's health standards for the six criteria pollutants are explained in the next sidebar. In addition, EPA and state environmental agencies monitor and regulate a wide range of other pollutants, usually referred to as "hazardous air pollutants" (HAPs). HAPs include benzene, 1,3-butadiene, hexavalent chromium, acrolein, and nearly two hundred other compounds. There are no federal standards limiting the amount of these pollutants found in ambient air. However, EPA does regulate emissions of these compounds under Title III of the Clean Air Act, which requires EPA to require industries to reduce these compounds using "maximum achievable control technology," or MACT. Dozens of HAPs are also regulated indirectly through Clean Air Act requirements to control ozone and PM, because many HAPs are volatile organic compounds (VOCs) and VOCs help form both ozone and particulate matter.

If pollutants are directly emitted, they are known as "primary pollutants." If they are formed in the atmosphere through chemical reactions of other pollutants, they are known as "secondary pollutants." CO, NO_2, SO_2, and lead are primary pollutants. Ozone is a secondary pollutant, formed from a series of reactions among oxides of nitrogen (NOx; refers to nitric oxide—NO—plus NO_2), VOCs, oxygen, and other compounds in the presence of sunlight. NOx and VOCs are therefore known as "ozone precursors."

PM is usually a mixture of primary and secondary pollutants. The secondary pollutants include ammonium nitrate, ammonium sulfate, and organic compounds. Ammonium nitrate is formed from gaseous ammonia and NOx. Ammonium sulfate is formed from ammonia and sulfur dioxide gas. Organic PM is formed from gaseous VOCs. Primary PM includes diesel soot, dust from roadways, construction, wind, and farming, industrial smoke emissions, and smoke from fireplaces and forest fires.

Pollution levels in the air we breathe (ambient levels) are directly measured by a network of more than a thousand monitoring sites positioned all around the United States. The criteria pollutants are the most widely monitored, but some states also measure various HAPs, as well as total VOCs. In addition, many states measure specific components of PM, such as sulfate, nitrate, and organic carbon.

As with the nation as a whole, these improvements are all the more impressive when one considers the enormous population growth in these high-pollution areas. The Los Angeles area achieved its extraordinary air quality improvements despite a 43 percent increase in the region's population between 1980 and 2000.

Air quality will continue to improve. Most pollution reductions required by existing regulations have not yet been realized. Fleet turnover to the progressively cleaner vehicles required by EPA's new-vehicle standards, which have been tightened several times over the last few decades, will reduce emissions by at least 80 percent during the next twenty years, even after accounting for growth in driving and the popularity of SUVs. Specifically, EPA's "Tier 2" regulation for automobiles, which began phase-in with the 2004 model year, will reduce per-mile emissions of the average automobile by about 90 percent over the next two decades. Similarly stringent requirements for heavy-duty diesel trucks phase in with the 2007 model year and in 2010 for off-road heavy diesel vehicles.

EPA's regulations for power plants and industrial boilers include the Clean Air Act's Title IV acid rain program, the NOx "SIP Call," the Clean Air Interstate Rule, and the Clean Air Mercury Rule. These regulations will reduce power plant emissions by 50–70 percent below current levels during the next two decades. EPA rules also continue to eliminate most emissions of dozens of other air pollutants, such as benzene, hexavalent chromium, polycyclic aromatic hydrocarbons, and acrolein, from motor vehicles, consumer products, and dozens of industries.

Regulators and environmental activists exaggerate air pollution levels and obscure positive trends. Regulators and environmentalists are the main purveyors of public information on air pollution, but the information they provide often has little to do with actual air pollution levels or trends. Both the EPA and the American Lung Association (ALA) claim that nearly half of all Americans breathe air that violates federal air pollution standards. In reality, about 11 percent of Americans live in areas that violate the 8-hour ozone standard, while about the same fraction live in areas that violate for $PM_{2.5}$.

EPA and ALA get their inflated numbers by counting *everyone* in a county as breathing air that exceeds federal standards, even if most of the

A Guide to Federal Air Pollution Standards

EPA has issued and periodically revised health standards for each of the criteria pollutants. These standards typically come in two forms—long-term standards based on annual-average pollution levels, and short-term standards based on levels averaged over a period of one to twenty-four hours. Table I-1 summarizes EPA's standards for criteria pollutants.

The "original" criteria pollutant standards have been around since the 1970s. In 1997, EPA created new and substantially more stringent standards for ozone and PM. For ozone, EPA added the 8-hour ozone standard to the preexisting 1-hour standard. For PM, EPA created an annual average and a 24-hour standard for "fine" particulate matter, or $PM_{2.5}$.

An annual pollution standard consists simply of an average pollution level that may not be exceeded. A region—usually a county or multicounty "air basin"—violates the standard if it has at least one monitoring location that violates the standard.

Short-term standards usually include not only a given "not-to-exceed" pollution level, but also a maximum frequency with which it can be exceeded while still staying in attainment of the standard.

For example, to violate the 1-hour ozone standard, a monitoring location must have at least four days during the most recent three years with a peak 1-hour ozone level greater than or equal to 0.125 parts per million (ppm). That is, to violate the standard, the monitoring location must have at least four exceedance days in a given three-year period.

To violate the 24-hour $PM_{2.5}$ standard, the average of the ninety-eighth percentile $PM_{2.5}$ values (roughly the seventh-highest daily average) from each of the last three years must be greater than or equal to 65.5 micrograms per cubic meter ($\mu g/m^3$). Once again, an entire county or air basin violates the standard if at least one monitor in the region is in violation.

For the 8-hour ozone standard, there isn't a one-to-one relationship between number of exceedances and violation of the standard, because it is not based on exceedances but on peak ozone concentrations. A monitoring location violates the 8-hour ozone standard if the average of the fourth-highest daily readings from each of the last three years is greater than or equal to 0.085 ppm. In practice, this means that most locations are in violation if they average four or more 8-hour exceedance days per year. However, some areas that violate the standard average only about three exceedances per year, while some attain the standard even with an average of five or six exceedances per year.

Thus, for short-term standards, an *exceedance* refers to a day in which pollution at a given monitoring location is greater than the nominal level

(continued on next page)

(continued from the previous page)

of the standard, but this does not necessarily mean that the monitor has *violated* a pollution standard. We maintain this distinction between an "exceedance" and a "violation" throughout this book.

Table I-1
Summary of Federal Air Pollution Standards

Pollutant	Standard	Averaging Times
Ozone (O$_3$)	0.085 ppm	8-hour[a]
	0.125 ppm	1-hour[b]
Particulate Matter (PM$_{2.5}$)	15 µg/m^3	Annual average[c]
	65 µg/m^3	24-hour[d]
Particulate Matter (PM$_{10}$)	50 µg/m^3	Annual average[e]
	150 µg/m^3	24-hour[f]
Carbon Monoxide (CO)	9 ppm	8-hour[g]
	35 ppm	1-hour[g]
Lead (Pb)	1.5 µg/m^3	Quarterly average
Nitrogen Dioxide (NO$_2$)	0.053 ppm	Annual average
Sulfur Dioxide (SO$_2$)	0.03 ppm	Annual average
	0.14 ppm	24-hour[g]

SOURCE: U.S. Environmental Protection Agency, "National Ambient Air Quality Standards," http://epa.gov/air/criteria.html (accessed November 27, 2006).

NOTES:
a. To attain this standard, the three-year average of the fourth-highest daily maximum 8-hour average ozone concentration must be less than 0.085 ppm. In May 2007, EPA proposed lowering the 8-hour ozone standard to somewhere between 0.060 ppm and 0.080 ppm. EPA will likely make a final decision by the end of 2007.
b. The standard is attained when the expected number of days per calendar year with maximum hourly average concentrations at or above 0.125 ppm is less than or equal to one. EPA revoked the 1-hour ozone standard in June 2005, except in fourteen metropolitan areas that violate the 8-hour ozone standard and that have entered into "early action compacts" with EPA, meaning they've agreed to take steps to attain the 8-hour ozone standard on a faster schedule than required under federal law.
c. To attain this standard, the three-year average of annual PM$_{2.5}$ concentrations must not exceed 15 µg/m^3.
d. To attain this standard, the three-year average of the ninety-eighth percentile of 24-hour PM$_{2.5}$ concentrations must not exceed 65 µg/m^3. In September 2006, EPA finalized a regulation reducing this standard to 35 µg/m^3, but enforcement of the new standard will not begin until 2010.
e. To attain this standard, the three-year average of annual PM$_{10}$ concentrations must not exceed 50 µg/m^3. EPA revoked this standard in September 2006.
f. Not to be exceeded on more than three days in any consecutive three-year period.
g. Not to be exceeded more than once per year.

county has clean air. For example, only one rural area of San Diego County, with about 1 percent of the population, violates EPA's 8-hour ozone standard. But EPA and the ALA count all three million people in the county as breathing "unhealthy" air. This is akin to giving every student in a school a failing grade if just one gets an "F" on an exam.

Chapters 5 and 6 show that these misleading portrayals are the norm in public information on air pollution, and provide realistic estimates of pollution levels where Americans live and breathe.

Air pollution affects far fewer people, far less often, and with far less severity than is commonly believed. The common wisdom is that a substantial fraction of Americans are suffering serious and permanent harm from air pollution even at recent historically low levels. Yet the actual research, most of it government-funded, tells a different story. California's Children's Health Study (CHS), for instance, reported that higher air pollution was associated with a *lower* risk of developing asthma.

The CHS also found that ozone had no effect on children's lung development, even in areas that exceeded the federal standard more than one hundred days per year. High levels of $PM_{2.5}$ were associated with a decline in lung function. However, even for children who grew up in areas with $PM_{2.5}$ at three times the level of the annual standard, the decline was only one to two percent. The CHS children were born in the early 1980s. No American child today experiences anywhere near the $PM_{2.5}$ levels that were associated with even these tiny lung-function declines.

Based on estimates by the California Air Resources Board, perhaps the most aggressive air pollution regulatory agency in the world, eliminating virtually all human-caused ozone in California would reduce total asthma-related emergency room visits and respiratory hospital admissions by only about 1 and 2 percent, respectively—and this in a state where millions of people live in areas that have by far the highest ozone levels in the nation. We show that even this estimate is inflated, because it is based on a selective reading of the health effects literature that ignores contrary evidence.

The most serious claim about air pollution is that it prematurely kills tens of thousands of Americans each year. The claim is based on small statistical correlations between daily pollution levels and daily deaths. But correlation doesn't necessarily mean causation, as recent embarrassing medical reversals have shown. For example, based on correlation studies, medical experts presumed that hormone-replacement therapy and Vitamin A supplements prevent heart disease, calcium supplements prevent osteoporosis, and a low-fat diet reduces cancer risks. But randomized trials, which eliminate the sources of uncontrollable bias that plague correlation studies, showed that these claims are greatly exaggerated or just plain false. In fact, Vitamin A turned out to *increase* cardiovascular risk.

The air pollution–mortality claim deserves even greater skepticism. First, it is based on the same unreliable correlation methods that have led medical authorities astray in other areas. Second, even though pollution was correlated with higher premature mortality on average, it seemed to *protect* against death in about one-third of cities. How could pollution kill people in some cities and save them in others? More likely, both results are chance correlations rather than real effects. Third, researchers have been unable to kill animals in laboratory experiments, even when they expose them to air pollution at levels many times greater than ever occur in the United States. This suggests that air pollution at today's record-low levels doesn't pose a risk, and current standards are health-protective with plenty of room to spare.

Conclusion

The book concludes with an analysis of how conflicts of interest inherent in federal air regulation have resulted in a regulatory system that harms the people it claims to be helping. The public's interest lies in sufficiently clean air, achieved at the lowest possible cost. But federal air quality regulation suffers from incentives to create requirements that are unnecessarily stringent, intrusive, bureaucratic, and costly.

The Clean Air Act charges the EPA with setting air pollution health standards. But this means that federal regulators decide when their own jobs are finished. Not surprisingly, no matter how clean the air, the EPA continues to find unacceptable risks. The EPA and state regulators' powers and budgets, as well as those of environmentalists, depend on a continued public perception that there is a serious problem to solve. Yet regulators are also major funders of the health research intended to demonstrate the need for more regulation. They also provide millions of dollars a year to environmental groups, which use the money to augment public fear of pollution and seek increases in regulators' powers. These conflicts of interest largely explain the ubiquitous exaggeration of air pollution levels and risks, even as air quality has steadily improved.

Exaggerating harm from air pollution makes us worse off overall. The public's interest is in an accurate portrayal of risk. Environmental regulations are not free. We have many needs and aspirations and scarce resources with which to fulfill them. When we devote excessive resources to one exaggerated risk, we give up opportunities to address other real and substantial risks, or to devote our hard-earned dollars to other things that would improve our lives, such as better education and health care, more nutritious food, bigger houses, and more leisure time.

Regulators' and environmentalists' power over Americans' lives continues to expand as EPA adopts ever more stringent standards. Americans will pay much to achieve these standards, but because our air is already safe to breathe, they will gain little in return. We show how results-focused regulatory policies can maintain healthful air, while avoiding the collateral damage caused by the perverse incentives inherent in the current system.

1

Air Quality Trends Before and After the Clean Air Act of 1970

For most of the twentieth century, air pollution control was mainly a state and local matter. That changed with the passage of the Clean Air Act Amendments of 1970 and the creation of the Environmental Protection Agency that same year. The Clean Air Act (CAA) nationalized air pollution control and began the process of developing a national regulatory regime to reduce air pollution.

At the time, proponents of federal control argued that state and local governments, competing for economic development, had little incentive to control air pollution and instead engaged in a "race to the bottom" of environmental quality in an effort to attract industry.[1] As a result, pollution increased throughout the twentieth century until the federal government stepped in and succeeded in controlling it where the states had shirked their duty.

The only problem with this mythology is that it isn't true. Contrary to the conventional wisdom, air quality improvement began decades before the CAA's passage. Far from being a story of "market failure," or a "race to the bottom" in environmental standards among states competing to attract jobs, the long-term history of air quality shows a "race to the top" of quality of life.[2]

Although systematic national data on air pollution are unavailable before the 1970s, a number of local and regional sources provide a patchwork history of air quality trends from the early 1900s onward. These data show that ambient levels of each type of air pollution were on the decline for as far back as the particular pollutant has been measured.[3]

Before the Clean Air Act:
Steady Improvements in Air Quality

Pittsburgh provides a key example of early trends in improving air quality. Photos of America's smokiest city on particularly smoky days at the beginning of the twentieth century appear to have been taken in the middle of the night.[4] Yet, as figure 1-1 shows, Pittsburgh reduced "dustfall"—an early measure of airborne particulate levels—by 50 percent between the 1920s and the 1940s, and halved it again between the 1940s and 1970. Other instances of improvement can be enumerated as well:

- Besides the statistics on "dustfall," figure 1-1 also shows more recent data based on airborne measurements of total suspended particulates (TSP)—that is, all airborne particulate matter—in Pittsburgh, as well as TSP averages for a large number of monitoring sites around the nation. TSP in Pittsburgh declined about 50 percent between the late 1950s and 1970.

- Figure 1-2 on page 16, from a 1970 paper by researchers from the National Air Pollution Control Administration—the forerunner of EPA's Office of Air and Radiation—shows that America's eastern and midwestern industrial cities achieved large reductions in particulate levels during the early and middle decades of the twentieth century.[5]

- Los Angeles County tripled its population between 1920 and 1940, with motor vehicle registrations rising by 40 percent between 1930 and 1940.[6] As a result of increasing motor-vehicle emissions, Los Angeles began to experience a new kind of air pollution in the late 1940s. This "photochemical haze" became known as smog and was composed of ozone and other irritating gases. Yet monitoring data from the 1950s and '60s show that by 1956, ozone had already begun a steady decline that continued through the 1960s and beyond (see figure 1-3 on page 17). The ozone decline may have begun even earlier, but data are not available before 1956.

FIGURE 1-1

LONG-TERM TRENDS IN PARTICULATE MATTER IN PITTSBURGH

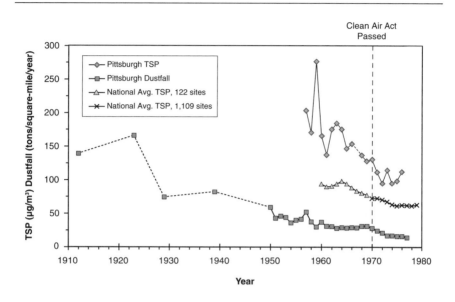

SOURCES: Pittsburgh data are from C. I. Davidson, "Air Pollution in Pittsburgh: A Historical Perspective," *Journal of the Air Pollution Control Association* 29 (1979): 1035–41. National data are from H. W. Ellsaesser, "Trends in Air Pollution in the United States," in *The State of Humanity*, ed. J. L. Simon (Malden, Mass.: Blackwell, 1995), 491–502, who cites EPA air quality trends reports from 1977 and 1984 as the ultimate sources.[7]

NOTES: TSP stands for "total suspended particulates," which was how PM was measured until regulators shifted their focus to PM_{10} in the mid-1980s. TSP is roughly equivalent to PM_{30}. Dustfall is the amount of particulate matter settling out of the air over a given period of time, and was measured by putting collection boxes on the roofs of buildings. "National Avg." is the average of all sites operating during the given time period. Dashed connecting lines mark year ranges in which data were not available for the intervening years.

Just as with smoke, once people perceived oxidant pollution as a problem, they quickly began to take action to mitigate it. Similarly, Americans began reducing other forms of air pollution years before the 1970 adoption of the Clean Air Act and the creation of EPA. For example, data from New York City show a 58 percent decline in sulfur dioxide levels in the seven years before the act's passage.[8]

A network of twenty-one national sulfur dioxide monitors likewise showed a 36 percent decline in average sulfur dioxide levels during the

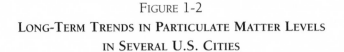

FIGURE 1-2

LONG-TERM TRENDS IN PARTICULATE MATTER LEVELS
IN SEVERAL U.S. CITIES

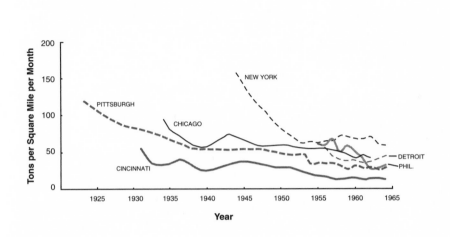

SOURCE: J. H. Ludwig, G. B. Morgan, and T. B. McMullen, "Trends in Urban Air Quality," *EOS* 51 (1970): 468–75.

NOTES: PM is measured here as dustfall—the amount of particulate matter falling on a given area over a given amount of time, measured in tons per square mile per month.

1960s (see figure 1-4 on page 18). These are the earliest sulfur dioxide data we were able to find. Declines were likely occurring even before this. The smokiness of eastern and midwestern American cities during the early to mid-1900s was associated with the burning of coal, which is also the source of the vast majority of sulfur dioxide emissions. The same processes that reduced smoke from coal-burning during the early 1900s to the 1960s likely also reduced sulfur dioxide.

Why Did Air Quality Improve *Before* the Clean Air Act?

The data show that ambient levels of the types of pollution people were aware of and concerned about were in decline long before the Clean Air Act

FIGURE 1-3

TREND IN DAYS PER YEAR EXCEEDING AN OZONE LEVEL OF 0.1 PPM IN LOS ANGELES, 1956–78

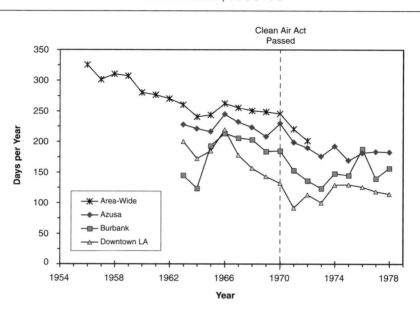

SOURCES: "Area-wide" data are from H. W. Ellsaesser, "Trends in Air Pollution in the United States," in *The State of Humanity*, ed. J. L. Simon (Malden, Mass.: Blackwell, 1995), 491–502. Data for specific monitoring sites were provided by staff at the California Air Resources Board.

NOTES: The "area-wide" value is the number of days per year in which at least one monitor in the Los Angeles region exceeded an oxidant level of 0.1 ppm. In order to check trends after the passage of the Clean Air Act, we obtained hourly oxidant monitoring data for all California sites from 1963 to 1978 from the California Air Resources Board (CARB). 1963 was the earliest year for which CARB had data in electronic form. Azusa, Burbank, and downtown Los Angeles were the only sites in the Los Angeles region with complete data for all years from 1963 through 1978. The data analysis methods are detailed in the note.[9]

was adopted in 1970. But this isn't the full story, because ambient data don't necessarily show when environmental cleanup began. For example, total pollution emissions might continue to grow due to a growing economy and a growing population, even if emissions per capita or per unit of economic activity are in decline. The cleanup also need not be due to any conscious effort, but could occur unintentionally due to market pressures for increased efficiency, technological progress, which both causes and is caused by economic growth, and changes over time in the types of goods

FIGURE 1-4

**NATIONAL TREND IN AVERAGE ANNUAL SULFUR DIOXIDE (SO₂)
LEVELS, 1962–86**

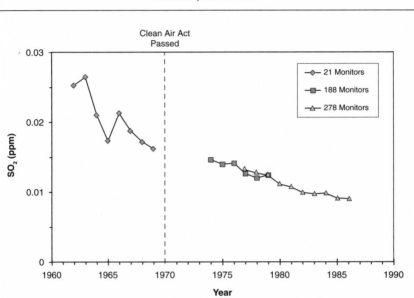

SOURCES: Data are from Bureau of the Census, *Statistical Abstract of the United States* (Washington, D.C.: U.S. Department of Commerce, 1981); Council on Environmental Quality, *Environmental Quality* (Washington, D.C.: Government Printing Office, 1971); Environmental Protection Agency, *National Air Quality and Emission Trends Report, 1976* (Washington, D.C.: Government Printing Office, 1977). Chart is adapted from I. M. Goklany, *Clearing the Air: The Real Story of the War on Air Pollution* (Washington, D.C.: Cato, 1999).

and services consumers need or want. This appears to be exactly what happened in America during the twentieth century.

One leading indicator of early environmental improvement is the ratio of emissions (E) per dollar of gross domestic product (GDP), expressed as E/GDP. Decades before they began to be perceived as air quality problems that needed to be solved, E/GDP for SO_2, NOx, VOC, and CO all began to decline.[10] For example, E/GDP for SO_2 began declining in the 1920s, but was not perceived to be a pollutant until around 1950. For NOx and VOC, E/GDP began to decline in the 1930s, but they were not recognized as pollutants until the 1950s, either.[11] The declines in E/GDP between these two time periods were quite large—more than 50 percent for SO_2 and VOC, and about 25 percent for NOx. These improvements occurred even though federal regulation of these pollutants did not begin until the late 1960s or early 1970s, and little state regulation occurred before the 1950s.

Regardless of whether the Clean Air Act has caused continued reductions in air pollution since 1970, the pre-CAA record shows that the "race to the bottom" justification for the federal takeover of air pollution regulation was false. Through a combination of common-law nuisance suits and local regulation, Americans had been addressing air pollution issues, as they were understood at the time, for several decades before the act was passed.[12]

Furthermore, throughout the twentieth century, market forces promoting greater efficiency and technological advancement often had the salutary side effect of reducing the pollution emitted per unit of economic activity, thereby moderating the air quality impacts of ongoing economic development.[13] For example, growing affluence allowed households to switch from coal to cleaner, more efficient natural gas for home heating and cooking. Railroads switched from coal-fired steam locomotives to diesel. The adoption of alternating current and improvements in transformer technology allowed power plants to be located near coal mines rather than in cities, because electricity could now be efficiently transported via long-distance power lines. The market forces that caused these transformations were not driven by air quality concerns, but they nevertheless caused large declines in air pollution levels.

Improvements in air quality are not unique. Other environmental problems, such as water quality, were also improving before the federal government took over regulatory control.[14] Likewise, other risks were dropping without federal regulation. Per mile of driving, the risk of dying in a car accident dropped 75 percent between 1925 and 1966—the year Congress adopted the National Traffic and Motor Vehicle Safety Act and created the National Highway Traffic Safety Administration.[15] Between 1930 and 1971—the year that the Occupational Safety and Health Administration was created—the risk of dying in a workplace accident dropped nearly 55 percent.[16] In all these cases—air pollution, automobile safety, and workplace safety—the rate of improvement was about the same before and after the federal government nationalized policy. Without doubt, improvements would have continued in all these areas even if the federal government had not taken the regulatory reigns away from the states.[17]

More Driving, More Energy, More Economic Activity . . .
and Less Air Pollution

Air pollution continued to decline apace after passage of the CAA, with virtually the entire country attaining federal standards for nitrogen dioxide, sulfur dioxide, and lead by the mid-1980s, and for carbon monoxide by the mid-1990s. Today only a few percent of monitoring locations still violate federal standards for particulate matter under ten microns in diameter (PM_{10}). Even for ozone, the most intransigent of the pollutants regulated under the CAA, the percentage of monitors failing EPA's original "1-hour" ozone standard went from 80 percent in the late 1970s to about 6 percent at the end of 2005, while the national-average number of days per year exceeding the standard declined more than 95 percent.

Lead was essentially eliminated as an air pollutant once most lead was removed from gasoline during the 1980s.[18] According to EPA, average ambient lead levels declined 96 percent between 1980 and 2005.[19] Lead levels probably peaked during the 1970s, due to a combination of rising automobile use and relatively high levels of lead in gasoline. Since the mid-1970s, ambient lead levels have declined by about 99 percent.[20]

EPA adopted more stringent standards for ozone and particulate matter in 1997. The newer ozone standard is based on daily 8-hour averages rather than 1-hour averages and is therefore known as the 8-hour ozone standard. Due to concerns that the finer fractions of particulate matter might be the most pertinent to health, the particulate matter standard focuses on $PM_{2.5}$—particulate matter under 2.5 micrometers in diameter—rather than PM_{10}.

Because they are more stringent, violation of these standards is more widespread. But here, too, the nation has made great progress. Where about 80 percent of ozone monitors violated the 8-hour standard in the late 1970s, only 15 percent violate the standard today, and the average number of days per year exceeding the standard has dropped more than 80 percent.[21] Likewise, the violation rate for $PM_{2.5}$ dropped from about 90 percent in the early 1980s to 14 percent by the end of 2006,[22] and average $PM_{2.5}$ levels declined 44 percent.

These improvements are all the more impressive when one considers that Americans have progressively done more of the things that would be expected to increase air pollution. For example, from 1980 to 2005, total automobile miles driven increased 95 percent, while heavy-duty diesel truck miles increased 109 percent.[23] The amount of coal burned for electricity increased 61 percent, while GDP increased 114 percent.[24] Nevertheless, air pollution of all kinds sharply declined (recall figure 1 in the introduction on page 4).

In the remainder of this chapter, we briefly summarize national trends in ambient nitrogen dioxide, sulfur dioxide, and carbon monoxide, as well as trends in other air pollutants for which EPA does not have specific ambient standards. Chapters 2 and 3 then provide more detailed information on ozone and particulate matter.

Trends in Nitrogen Dioxide, Sulfur Dioxide, and Carbon Monoxide

Figure 1-5 on page 22 displays the national-average trend in NO_2 and SO_2 levels, while figure 1-6 on page 23 displays CO levels. In both charts the data are shown for all monitors that happened to be operating in a given year, as well as a twenty-year trend for those that operated continuously from 1984 to 2004 (dashed lines). Average levels of all three pollutants have steadily declined. During the last twenty years, NO_2 declined 31 percent, SO_2, 61 percent, and CO, 70 percent. The continuous data show that these declines are real and not due merely to changes in the number or location of the monitoring sites.

Figure 1-7 on page 24 displays the trend in the fraction of monitoring sites violating the federal standard for each pollutant. EPA has more than one standard for CO and SO_2, so we provide data for the standard with the highest violation rate. As the graph shows, virtually the entire nation has complied since the early 1990s with these standards for CO and since the early 1980s for NO_2 and SO_2.

Hazardous Air Pollutants

There are only six so-called "criteria" air pollutants—those for which EPA has set specific limits on allowable ambient levels. But EPA regulates air

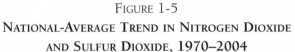

FIGURE 1-5

NATIONAL-AVERAGE TREND IN NITROGEN DIOXIDE
AND SULFUR DIOXIDE, 1970–2004

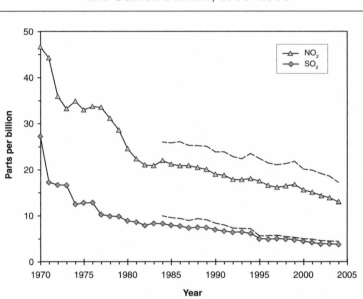

SOURCE: Analysis of monitoring site data downloaded from EPA's Air Quality System (AQS) database at http://www.epa.gov/ttn/airs/airsaqs/detaildata/downloadaqsdata.htm and http://www.epa.gov/ttn/airs/airsaqs/archived%20data/downloadaqsdata-o.htm (accessed March 21, 2006).

NOTES: The 1970–2004 trend includes all monitoring locations that happened to be operating in a given year, while the dashed lines provide the twenty-year trend for all sites that operated continuously from 1984 to 2004. Trends are based on average annual pollutant levels at each monitoring site

emissions of nearly two hundred other chemicals that are usually referred to as Hazardous Air Pollutants, or HAPs. Regulations that reduce VOC emissions in order to control ozone and PM also reduce HAPs, because many HAPs are VOCs.

California has collected trend data on a number of HAPs since 1990. Figure 1-8 on page 25 displays the trends in ambient levels of four key HAPs: benzene, 1,3-butadiene, hexavalent chromium (chromium VI), and benzo(a)pyrene. The pollutant levels in the graph are annual averages constructed by averaging levels from all sites that had valid data for all years included in the graph.[25] Since most monitoring sites were missing valid

FIGURE 1-6

NATIONAL-AVERAGE TREND IN CARBON MONOXIDE, 1970–2004

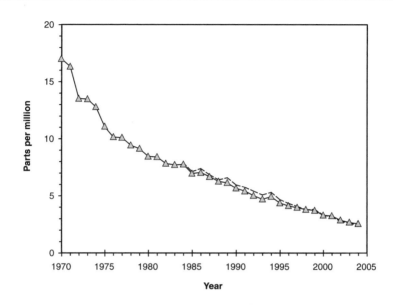

SOURCES: Analysis of monitoring site data downloaded from EPA's Air Quality System (AQS) database at http://www.epa.gov/ttn/airs/airsaqs/detaildata/downloadaqsdata.htm and http://www.epa.gov/ttn/airs/air-saqs/archived%20data/downloadaqsdata-o.htm (accessed March 21, 2006).
NOTES: The 1970–2004 trend includes all monitoring locations that happened to be operating in a given year, while the dashed line provides the twenty-year trend for all sites that operated continuously from 1984 to 2004. Trends are based on the second-highest 8-hour-average CO level measured each year at each monitoring location.

annual averages for some years, we chose sites that maximized geographic representation and the number of years with complete data.[26]

Benzene and 1,3-butadiene are emitted mainly by gasoline-fueled motor vehicles.[27] Despite an estimated 2.5 percent per year increase in total vehicle miles traveled in California, ambient levels of these pollutants have been dropping about 11 percent and 9 percent per year, respectively.[28]

Benzo(a)pyrene is one of a family of organic compounds known as polycyclic aromatic hydrocarbons (PAH), which are components of soot. Benzo(a)pyrene and other PAHs are produced by diesel engines, other

FIGURE 1-7

PERCENTAGE OF U.S. MONITORING LOCATIONS VIOLATING THE FEDERAL
NO₂, SO₂, AND CO STANDARDS, 1975–2004

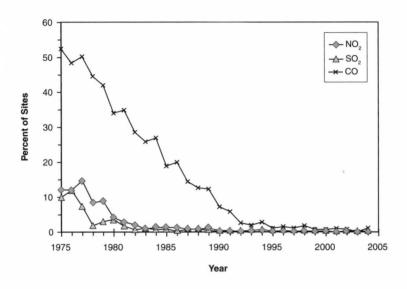

SOURCE: Analysis of monitoring site data downloaded from EPA's Air Quality System (AQS) database at http://www.epa.gov/ttn/airs/airsaqs/detaildata/downloadaqsdata.htm and http://www.epa.gov/ttn/airs/airsaqs/archived%20data/downloadaqsdata-o.htm (accessed March 21, 2006).

NOTES: For each pollutant, chart is based on the standard with the highest violation rate (for example, the 8-hour CO standard, rather than the 1-hour CO standard).

fossil-fuel combustion sources, and forest fires. Benzo(a)pyrene levels have been dropping an average of 8 percent per year. The California Department of Transportation estimates that diesel fuel use has been increasing about 2 percent per year during the last few years. Nevertheless, benzo(a)pyrene levels have been rapidly declining.[29] Hexavalent chromium is emitted mainly by metal-plating facilities. Hexavalent chromium levels have been dropping by 9 percent per year. Data from other areas of the United States demonstrate similar reductions in a wide range of HAPs during the last fifteen years, once again despite substantial increases in population, economic activity, energy production, and vehicle-miles of driving.[30]

FIGURE 1-8
TRENDS IN AMBIENT LEVELS OF FOUR HAPS AT CALIFORNIA
MONITORING LOCATIONS, 1990–2005

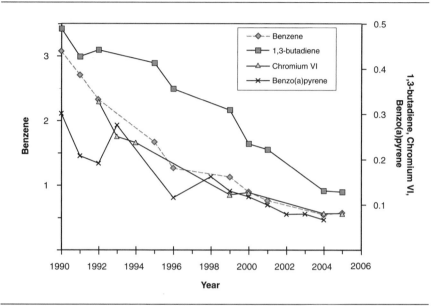

SOURCE: CARB annual toxics monitoring data retrieved from California Air Resources Board, *2007 Air Pollution Data CD*, January 2007, http://www.arb.ca.gov/aqd/aqdcd/aqdcd.htm (accessed March 21, 2007).

NOTES: Benzene and 1,3-butadiene are in parts per billion. Chromium VI and benzo(a)pyrene are in nanograms per cubic meter.

Conclusion

Air pollution has been declining for as far back as we have measurements—since the early twentieth century for some cities. Early improvements resulted from a combination of market forces and common-law nuisance suits, with local and state regulation playing an increasing role as the century progressed. Americans sharply reduced air pollution in their communities long before the federal Clean Air Act nationalized pollution control in 1970. These improvements have continued since 1970, with virtually the entire nation attaining federal standards for four of the six criteria pollutants and greatly reducing levels of additional pollutants for which EPA has not set specific ambient standards. Ozone and $PM_{2.5}$ are the only remaining air pollutants for which nonattainment of federal standards is relatively common. Nevertheless, as we show in the next two chapters, levels of these pollutants also continue to decline.

2

Ozone: Historic Trends and Current Conditions

Ozone levels have declined substantially since the 1970s, with the worst areas experiencing the largest drops. As of the end of 2006, 94 percent of U.S. monitoring locations complied with the federal 1-hour ozone standard, up from about a 20 percent attainment rate during the late 1970s. At the end of 2006, 85 percent of monitors complied with the 8-hour standard, also up from 20 percent in the late 1970s.

The national-average number of days per year exceeding ozone standards likewise declined between 1975 and 2006—by more than 95 percent for the 1-hour standard, and more than 85 percent for the 8-hour standard. The 2003–6 period had the lowest ozone levels since national monitoring began in the 1970s.

Although levels are improving, the 8-hour ozone standard is nevertheless the most widely violated of EPA's air pollution standards.[1] Furthermore, despite overall improvements, and despite measured decreases in emissions of ozone-forming pollutants, 8-hour ozone levels declined only slightly during the 1990s, and even increased in a few places, before dropping during the last few years. The complications of the chemistry of ozone formation, combined with regulators' imprudent choices on ozone control strategies, might explain these variable trends.

Although ozone levels are relatively low around the United States, it remains the most stubborn of the nation's air pollution challenges. Large reductions in ozone "precursor" pollutants have only a modest effect on ozone levels. This chapter explores ozone trends and levels and the complications of ozone chemistry, and observes that the nation's worst ozone problems are limited to a few parts of California.

National Trends in Ozone Levels

Ozone levels have been declining for decades around the United States. Figure 2-1 shows the trend in the average number of days per year exceeding the federal 1-hour and 8-hour ozone standards during the last thirty years. The solid lines are based on data from all monitoring sites that happened to be operating in a given year. However, this could obscure real trends, because some sites go into and out of operation over time. We therefore also present trends only for 253 monitors that operated continuously from 1985 to 2006, as shown by the broken lines.

Note that the shape of the continuous-sites trend is similar to that of the all-sites trend, demonstrating that the ozone declines are real. In fact, the continuous-sites trend is even steeper than the all-sites trend. Although the continuous sites, on average, now have ozone levels similar to all sites, note that in the past they had significantly more ozone exceedance days. This is to be expected, as areas with the worst air pollution have naturally tended to have the most intensive pollution monitoring.

In any case, figure 2-1 shows that the nation has achieved extraordinary declines in peak ozone levels over the last thirty years. The national-average number of days per year exceeding the 1-hour standard declined more than 95 percent between 1975 and 2006, while 8-hour exceedance days declined more than 85 percent. Eight-hour ozone levels have declined in fits and starts with large year-to-year variations, while 1-hour levels have gone down relatively steadily. The last four years are the four lowest ozone years on record.

The large variation in ozone exceedances is due mainly to year-to-year variations in weather. All else equal, ozone will be higher in warmer, sunnier years with low winds and low rainfall. Superimposed on these year-to-year variations is a long-term decline in ozone levels, due to declining emissions of ozone-forming pollutants (VOCs, NOx). The years 2005 and 2006 were among the hottest on record, which favors higher ozone levels, yet these two years were also the second- and third-lowest on record for ozone.[2]

Figure 2-2 on page 30 shows the trend in the percentage of national monitoring locations violating the 1-hour and 8-hour ozone standards, once again for all sites and for continuous sites. The trend in the violation

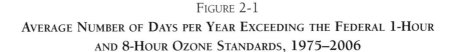

FIGURE 2-1

AVERAGE NUMBER OF DAYS PER YEAR EXCEEDING THE FEDERAL 1-HOUR
AND 8-HOUR OZONE STANDARDS, 1975–2006

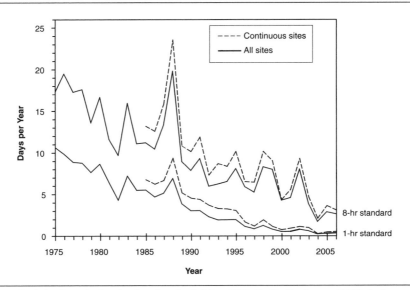

SOURCE: Analysis of monitoring site data downloaded from EPA's Air Quality System (AQS) database at http://www.epa.gov/ttn/airs/airsaqs/detaildata/downloadaqsdata.htm, http://www.epa.gov/ttn/airs/airsaqs/archived%20data/downloadaqsdata-o.htm, and http://www.epa.gov/air/data/geosel.html (accessed April 8, 2007).

NOTES: Between 1975 and 2006, 8-hour ozone exceedance days declined more than 85 percent, while 1-hour exceedance days declined more than 95 percent. The graph displays the trend in ozone exceedances per year in two ways. "All sites" provides the trend from 1975 to 2006 for all ozone monitoring sites that happened to be operating in a given year. This could create bias, because some sites may be added or removed in any given year. Thus, we also show a trend from 1985 to 2006 for only those sites that operated continuously during that twenty-two-year period (253 sites). The two trends are highly correlated, showing that "all sites" provides a valid representation of the trend in ozone exceedances. If anything, the "all sites" trend understates the decline in ozone levels over time, as the "continuous sites" trend has a steeper downward slope.

rate doesn't drop as steadily as the trend in exceedance days, particularly for the 8-hour standard. This is because violation of an ozone standard is based on three consecutive years of data. Thus, one year with high ozone will affect the violation rate for three years. For example, the uniquely high levels of 1988 (see figure 2-1) caused the violation rate to be high not only in 1988 but also in 1989 and 1990 (see figure 2-2 on the following page).

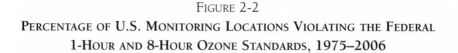

FIGURE 2-2

PERCENTAGE OF U.S. MONITORING LOCATIONS VIOLATING THE FEDERAL
1-HOUR AND 8-HOUR OZONE STANDARDS, 1975–2006

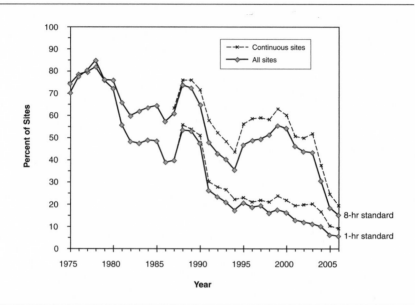

SOURCES: Analysis of monitoring site data downloaded from EPA's Air Quality System (AQS) database at
http://www.epa.gov/ttn/airs/airsaqs/detaildata/downloadaqsdata.htm, http://www.epa.gov/ttn/airs/airsaqs/
archived%20data/downloadaqsdata-o.htm, and http://www.epa.gov/air/data/geosel.html (accessed April 8,
2007).

The 8-hour violation rate rose in the late 1990s due to consecutive high-
ozone years in 1998 and 1999. Likewise, the precipitous drop in the violation rate from 2003 to 2005 was due to unusually low ozone levels in 2004,
followed by the removal of 2002, a high-ozone year, from the calculation of
the ozone exceedance rate for 2005. The violation rate remained low in
2006 as low ozone levels continued for a fourth consecutive year.

We now turn to current ozone levels across the United States. The results
we present below are based on analysis of data from 1,035 monitoring
sites with valid data for 2004–6. Figure 2-3 on page 32 gives the range of
exceedance days per year in each state for the 8-hour and 1-hour standards.
The diamond marks the average level for all monitors in a given state, while
the vertical lines mark the range from worst to best monitoring site.

As of the end of 2006, twenty states had at least one location that violated the 8-hour ozone standard, and ten had at least one location that violated the 1-hour standard. With the exception of the worst areas of California, which have always been in a class by themselves, ozone exceedances are relatively rare throughout the United States. Most of the country never exceeds the 1-hour standard, and even the worst areas outside California now average only ten or fifteen 8-hour exceedance days per year.

Exceedances of the 1-hour ozone standard have actually been rare for a long time. Between 1995 and 2005, 68 percent of monitoring locations had no exceedance days, and another 24 percent averaged less than one per year. There have been a few years, for example 1998 and 2002, when 8-hour exceedances were relatively common in some areas outside California. For example, some eastern metropolitan areas experienced as many as a few dozen during 1998 and 2002 at their highest-ozone locations. In contrast, the worst areas of California have experienced between seventy and one hundred fifty 8-hour exceedance days per year during the last decade.

The appendix to this chapter on page 203 provides state-by-state data on 1-hour and 8-hour ozone trends, and shows that ozone declined not only on a national-average basis but also around the United States.

Figure 2-4 on page 33 shows the effect of a tighter ozone standard based on recent monitoring data. In June 2007, EPA proposed lowering the 8-hour standard from its current level of 0.085 ppm down to somewhere between 0.060 ppm and 0.080 ppm.[3]

As the graph shows, many metropolitan areas would currently be coming into attainment based on the current standard. However, even if EPA lowers the standard modestly, down to 0.080 ppm, 43 percent of the nation's metro areas would be out of attainment. The nonattainment fraction would double to 87 percent for a standard set at 0.070 ppm, which is a likely level for EPA to choose. In other words, EPA is planning a vast expansion of the reach of the Clean Air Act, one that would turn nearly the entire nation into a "nonattainment" area. For many areas, this non-attainment status will be effectively permanent, because a standard set at 0.070 ppm would likely be unattainable in much of the nation, even with a virtual elimination of all motor-vehicle and industrial emissions (see next section of this chapter for more on this).

FIGURE 2-3

RANGE OF DAYS PER YEAR EXCEEDING THE 8-HOUR (BOTTOM) AND 1-HOUR (TOP)
OZONE STANDARDS IN EACH STATE (PLUS THE DISTRICT OF COLUMBIA), 2004–6

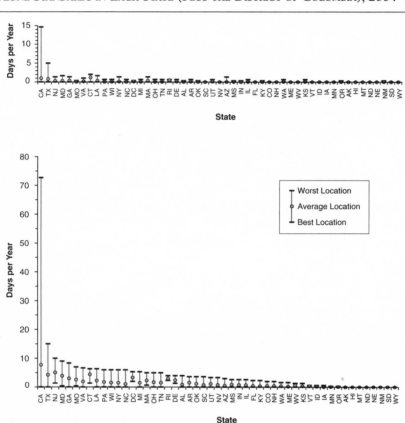

State

Source: Analysis of monitoring site data downloaded from EPA's AIRData database, http://www.epa.gov/air/ data/geosel.html (accessed March 21, 2007).

NOTES: The graphs were constructed by averaging the annual number of ozone exceedance days during 2004–6 at each site and then calculating the average and range of these values for each state.

Figure 2-5 on page 34 shows the extent to which a few California areas are in a class by themselves in terms of ozone pollution, and also the extent to which ozone can vary from place to place even within a county. The chart gives the range of 8-hour exceedance days per year in the forty counties with the worst ozone (based on the location in each county with the most

FIGURE 2-4

PERCENTAGE OF U.S. METROPOLITAN AREAS THAT WOULD VIOLATE A TOUGHER 8-HOUR OZONE STANDARD

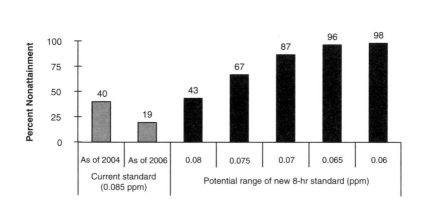

SOURCE: Analysis of monitoring site data downloaded from EPA's AIRData database, http://www.epa.gov/air/data/geosel.html (accessed March 21, 2007).

NOTES: The two left bars show the percentage of metropolitan areas that violated the 8-hour ozone standard based on data for 2002–4 ("as of 2004") and 2004–6 ("as of 2006"). In other words, many metropolitan areas would currently be coming into attainment based on the current standard. The next five bars show the percent of metro areas that would fail to attain a progressively tougher ozone standard.

exceedance days) during 2003–5. The diamond marks the average level for all monitors in a given county, while the vertical lines mark the range from worst to best monitoring site. Note that some counties have only one monitor and therefore only a single ozone value plotted in the chart.

First, note that several California counties have both worst and average ozone levels that are well above those of even the worst portions of all other counties in the United States. Second, note the wide range of exceedances within counties that have more than one ozone monitor. In several cases, even though the worst area of the county might have ten or more exceedance days per year, the best area has only a few or even zero exceedances. Both of these observations will come into play later in this book, when we discuss why public perception of air pollution is so different from reality.

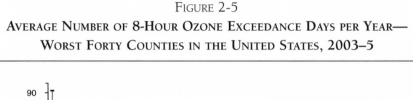

FIGURE 2-5

AVERAGE NUMBER OF 8-HOUR OZONE EXCEEDANCE DAYS PER YEAR—
WORST FORTY COUNTIES IN THE UNITED STATES, 2003–5

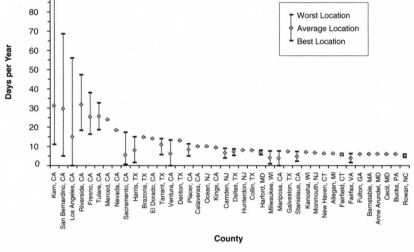

SOURCE: Analysis of monitoring site data downloaded from EPA's AIRData database, http://www.epa.gov/air/data/geosel.html (accessed September 29, 2006).

The Chemistry of Slower Progress on 8-Hour Ozone Levels

If there is a pessimistic note in all the good news on air quality trends, it is that 8-hour ozone levels have improved more slowly than those of other pollutants, declining only slightly during the 1990s before hitting the record lows of the last few years. The slow progress of the '90s occurred despite continued steady reductions in nitrogen oxides (NOx) and volatile organic compounds (VOC)—the two pollutants that help form ozone (see chapter 4 for detailed information on emissions trends).

The reasons for this are still a matter of scientific debate, but they likely relate to at least two factors. First, computer modeling of ozone chemistry suggests that it takes large reductions in ozone-forming emissions to achieve small reductions in ozone levels. For example, studies of ozone formation in California and the eastern United States suggest that reducing peak 8-hour levels by 10 or 20 percent—about the amount necessary to attain the current 8-hour standards in most areas—will require reductions in ozone-forming emissions on the order of 70–90 percent below late-1990s levels.[4] In the Los Angeles metro region, the current 8-hour standard might not be attainable even with a virtual elimination of all human-caused ozone-forming emissions.[5] EPA's plan to tighten the standard still further will make ozone attainment a virtual impossibility in much of the nation.

Second, the formation of ozone depends not only on the total amounts of ozone-forming pollutants in air, but also on their ratio. When the VOC-to-NOx ratio is low, VOC reductions reduce ozone, but NOx reductions can actually *increase* it, or at least mute the effect of the VOC reductions (see the sidebar in this chapter for more on how this works). This appears to be happening in many cities around the country and suggests that regulators may need to rethink their strategies for achieving the additional reductions necessary to attain the 8-hour standard.

A key source of evidence that NOx reductions can be detrimental to ozone control is a phenomenon known as the "weekend effect." NOx emissions decline by as much as 60 percent on weekends relative to weekdays, mainly due to much lower use of diesel vehicles on weekends. VOCs decline as well, but by substantially less, because people still do a lot of driving on weekends and most VOCs come from gasoline-powered vehicles and equipment. Despite these substantial reductions in the pollutants that form ozone, ozone levels stay the same or even *increase* on weekends in most urban areas around the nation. The phenomenon has been observed for decades in some areas, but monitoring data show that it is increasing in magnitude and has become the norm at metropolitan monitoring sites around the country.[6]

For example, in the Los Angeles region, NOx levels are about 25–40 percent lower on Sundays than on weekdays, but ozone levels are 20–50 percent higher.[7] Even though weekends account for only 29 percent of all days of the year, nearly 50 percent of 8-hour ozone exceedances in the Los Angeles metro area occur on weekends. In Atlanta, NOx drops nearly 60 percent and VOCs nearly 20 percent on Sundays relative to Wednesdays, yet ozone levels hardly budge.[8] In Cincinnati, NOx declines 40 percent on weekends relative to weekdays, but ozone rises slightly.[9]

In other words, natural ozone-control experiments are going on every weekend all around the United States. But large reductions in ozone-forming emissions are not reducing ozone and are often associated with ozone increases. A number of studies have provided evidence that NOx reductions are the culprit.[10]

We showed above that the last four years have had the lowest overall ozone levels on record. Thus, even though NOx reductions appear to be detrimental to ozone control in many areas, they have not prevented these declines. One explanation is the pattern of VOC and NOx emissions reductions. In metropolitan areas, VOCs have been declining more rapidly than NOx, due to large reductions in VOCs from automobiles—their major source—during the last decade, but smaller reductions in the major sources of metropolitan NOx—automobiles, diesel trucks, and off-road diesel vehicles (see chapter 4 for details on emissions trends).

In contrast to metropolitan areas, most NOx in rural areas comes from large, coal-fired power plants. But these power plants have reduced their ozone-season NOx emissions by about 60 percent since 1998 (once again, see chapter 4 for details). NOx reductions are more likely to be effective for reducing ozone in rural areas because of their higher VOC-to-NOx ratios relative to urban areas.

Thus, rural ozone may have been reduced due to power-plant NOx reductions. This would also reduce transport of rural ozone into urban areas, where most people live and where most pollution monitors are located. At the same time, ozone would have declined in metropolitan areas because rapid VOC reductions have been large enough to overcome the detrimental effects of NOx reductions in metropolitan areas. Or to put it another way, 8-hour ozone levels declined due to VOC reductions, but they might have declined even more if NOx had not been reduced in metropolitan areas.

This scenario is consistent with the evidence, but is by no means confirmed. Regardless, the fact remains that large reductions of ozone-forming emissions occur on weekends, but ozone levels nevertheless stay the same or increase. NOx reductions appear to be a major culprit. Whatever improvements in ozone have been achieved in recent years, it is likely that ozone would have been even lower if there had been fewer NOx reductions in metropolitan areas.

How Reducing NOx Can Make Ozone Worse

NOx (shorthand for NO_2 + NO) is necessary for the formation of ozone.[11] Even when VOCs are not present, ozone is formed from a series of reactions among NO_2, NO, and oxygen, and driven by sunlight:

$$NO_2 + sunlight \rightarrow NO + O \qquad (1)$$
$$O + O_2 \rightarrow O_3 \qquad (2)$$
$$O_3 + NO \rightarrow O_2 + NO_2 \qquad (3)$$

This series of reactions forms a cycle that results in a relatively small amount of ozone in the atmosphere. Ozone can't build up because it is also destroyed by reaction with NO to regenerate NO_2 (reaction (3)).

VOCs allow ozone to build up by creating a new route to regenerate NO_2 without destroying ozone. In other words, VOCs allow reaction (3), which destroys ozone, to be bypassed. OH radicals (also generated by various reactions among pollutants in the atmosphere) convert some VOCs to peroxy radicals, which then regenerate NO_2 as follows:

$$VOC - OO + NO \rightarrow NO_2 + VOC - O \qquad (4)$$

where the two oxygen atoms ("OO") are the peroxide group attached to a VOC.[12]

Ozone formation depends on the ratio of VOCs to NOx (VOC/NOx). At high VOC/NOx, ozone formation is controlled by the amount of NOx available, and reaction (4) is the main route to regenerate NO_2 from NO. Under this "NOx-limited" situation, decreasing NOx reduces ozone, while decreasing VOCs has little or no effect on ozone levels.

At low VOC/NOx, ozone formation is limited by the amount of VOCs available for reaction (4), and reaction (3) becomes the main route to regenerate NO_2 from NO. In addition, at low VOC/NOx, NO_2 competes with VOCs to react with OH radicals, slowing the rate at which VOCs are converted to peroxy radicals, and thereby slowing the rate of reaction (4).

Under this "VOC-limited" or "VOC-sensitive" condition, reducing VOCs reduces ozone, but reducing NOx *increases* ozone. NOx reductions increase ozone in two ways: first, by slowing down the rate of ozone destruction through reaction (3), and second, by speeding up the rate of NO_2 regeneration through reaction (4), thereby allowing each molecule of NOx to make ozone more rapidly.[13]

During the late 1990s, EPA embarked on a policy of seeking large NOx reductions in metropolitan areas. The resulting automobile, heavy-duty diesel truck, and nonroad diesel equipment standards will eliminate more than 80 percent of metropolitan NOx emissions from motor vehicles during the next two decades or so. These upcoming NOx reductions run a serious risk of slowing or even reversing progress on ozone. This problem will only become more acute if EPA follows through on its plan to adopt a tougher ozone standard. A less-risky policy would have been, and still could be, to focus on VOC reductions from mobile sources, while leaving mobile-source NOx emissions at 1990s levels, and focusing NOx-reduction efforts on rural NOx emissions, such as those from power plants.

Conclusion

Los Angeles was the first city to notice a photochemical or ozone smog problem, back in 1948. We showed in the previous chapter that Los Angeles began reducing ozone by at least the mid-1950s, and levels have been dropping there ever since. The rest of the nation did not begin monitoring ozone until the 1970s, but these data also show steady declines. Today, more than 80 percent of the nation complies with EPA's 8-hour ozone standard, and nearly 95 percent complies with the old 1-hour standard. Even areas that exceed the standards generally do so only a few days per year. Frequent exceedance of the 8-hour standard is limited mainly to just a few localized areas of California. Most of the country never exceeds the 1-hour standard at all.

3

Particulates: Historic Trends and Current Conditions

Airborne particulate matter (PM) is currently, along with ozone, the focus of intensifying regulatory effort. As with ozone, PM levels have been dropping for decades. Levels of fine particulate matter ($PM_{2.5}$) have declined more than 40 percent since the early 1980s and 15 percent since 1999, with similar improvements for PM_{10}. Only a few percent of monitoring locations still exceed federal PM_{10} standards, generally by small margins. Attainment is also virtually complete for EPA's current 24-hour $PM_{2.5}$ standard, with only one monitoring location failing to comply as of the end of 2006. About 13 percent of monitoring sites still violate the annual $PM_{2.5}$ standard, though the violation rate dropped by more than half between 2001 and 2006, and most violating sites are close to attainment of the standard.

Fifteen percent of the nation's metropolitan areas are in non-attainment under the current $PM_{2.5}$ standards. EPA recently finalized a more stringent 24-hour $PM_{2.5}$ standard that would increase the violation rate to 22 percent of metro areas, based on current $PM_{2.5}$ levels. EPA will begin enforcing this new standard in 2010.

National Trends in Particulate Matter (PM_{10} and $PM_{2.5}$)

As with other pollutants, levels of airborne particulate matter have been dropping for decades. National-average levels of particulate matter under 2.5 microns in diameter ($PM_{2.5}$) declined more than 40 percent between 1980 and 2006, despite a doubling of vehicle-miles of driving and more than a 60 percent increase in coal consumption. $PM_{2.5}$ dropped about

15 percent between 1999 and 2006. National-average PM_{10} declined 30 percent between 1990 and 2006.[1] Based on limited trend data from the early to mid-1900s (see chapter 1), PM levels in eastern and midwestern industrial cities have probably declined about 90 percent since their early twentieth-century peaks.

Virtually the entire nation attains the 24-hour standard for $PM_{2.5}$. However, about 13 percent of $PM_{2.5}$ monitors violated the annual standard as of the end of 2006.[2] In addition, in September 2006 EPA tightened the 24-hour $PM_{2.5}$ standard, from 65 $\mu g/m^3$ down to 35 $\mu g/m^3$, which will place about 24 percent of monitoring locations in violation (based on $PM_{2.5}$ data for 2004–6).[3]

The vast majority of the nation attains federal standards for particulate matter under 10 microns in diameter (PM_{10}). Although about 10 percent of monitoring locations violated the 24-hour and/or annual standards as of the end of 2004, the geographic extent of PM_{10} violation is much smaller than this percentage would suggest. After the violation rate had dropped to 2 percent by the mid-1990s, many monitoring sites in low-PM_{10} locations were retired, while additional monitors were added in the few remaining high-PM_{10} areas. The pattern of PM_{10} violation is different than for $PM_{2.5}$. Elevated $PM_{2.5}$ levels occur mainly in urbanized areas as a result of motor vehicle, power plant, and fireplace emissions, while high PM_{10} levels are largely a result of windblown dust in more rural areas.

The understanding and regulation of airborne particulate matter have steadily evolved over the last several decades. As discussed in chapter 1, PM was initially measured by the low-tech method of "dustfall" per unit area per unit time. By the 1950s and 1960s, technology had advanced to the point where particulate concentrations in air could be measured directly, leading to regulations based on "total suspended particulates" (TSP)—that is, all airborne PM.

As health research progressed, concern focused on smaller particles, as these appeared able to travel deeper into people's lungs. In 1987 EPA began regulating particulate matter under 10 microns in diameter rather than TSP (TSP is roughly equivalent to PM_{30}). By the mid-1990s, scientists had begun homing in on even finer fractions of PM as the main health culprit. Hence, in 1997, EPA adopted standards for $PM_{2.5}$—particulate matter under 2.5 microns in diameter—while also keeping the old PM_{10} standards in place.

Besides standards, methods of measuring airborne particulate levels have also evolved. When EPA adopted the new $PM_{2.5}$ standards, it also issued a new "federal reference method" (FRM) for $PM_{2.5}$ measurement. For any given actual ambient level, the new FRM gives a significantly higher $PM_{2.5}$ reading than previous methods, which must be accounted for when estimating long-term trends. Because EPA set its $PM_{2.5}$ standards based on data collected via the old measurement method, this also means that federal standards are effectively more stringent than they appeared to be when they were adopted.

Trends in $PM_{2.5}$

Widespread monitoring of $PM_{2.5}$ did not begin until 1999, but there are nevertheless some earlier data that can be used to assess long-term trends. From 1979 to 1984, EPA operated the Inhalable Particulate Network (IPN), which measured $PM_{2.5}$ levels in about ninety metropolitan areas around the nation.[4] Figure 3-1 on the following page displays the trend in the average $PM_{2.5}$ level for eighty-six IPN monitoring locations that also have measurements for 2000–6 and for 269 monitoring sites with continuous data for 1999–2006.

The trend from the eighty-six IPN sites matches almost perfectly with the more extensive data collected during the last several years. This confirms that IPN sites can be taken as a valid representation of national-average $PM_{2.5}$ levels during the early 1980s.

Between the early 1980s and 2006, national $PM_{2.5}$ levels declined an average of 44 percent.[5] The violation rate went from about 90 percent of sites in the early 1980s down to 13 percent in 2006.[6] All areas had lower $PM_{2.5}$ in 2006 than during the early 1980s. Areas with the worst $PM_{2.5}$ experienced the largest improvements, both in percentage reduction and absolute amount of reduction. For example, Riverside, California, went from 48 µg/m³ in the early 1980s down to 20 µg/m³ today—a reduction of 58 percent, or 28 µg/m³. Steubenville, Ohio, went from 32 µg/m³ in the early 1980s down to 14 µg/m³ today—a reduction of 56 percent, or 18 µg/m³.

$PM_{2.5}$ has continued to decline more recently as well. National-average $PM_{2.5}$ dropped about 15 percent between 1999 and 2006, and the fraction

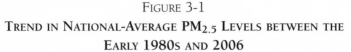

FIGURE 3-1

TREND IN NATIONAL-AVERAGE PM₂.₅ LEVELS BETWEEN THE
EARLY 1980S AND 2006

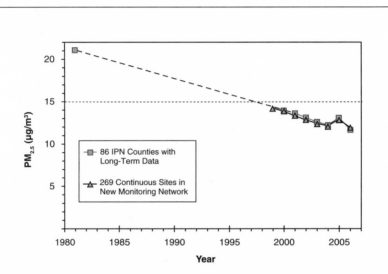

SOURCES: IPN data were retrieved from U.S. Environmental Protection Agency, *Inhalable Particulate Network Report: Operation and Data Summary (Mass Concentrations Only)*, Volume I, April 1979–December 1982, by D. O. Hinton, J. M. Sune, J. C. Suggs, and W. F. Barnard (Washington, D.C.: Government Printing Office, November 1984), and U.S. Environmental Protection Agency, *Inhalable Particulate Network Report: Data Summary (Mass Concentrations Only)*, Volume III, January 1983–December 1984, by D. O. Hinton, J. M. Sune, J. C. Suggs, and W. F. Barnard (Washington, D.C.: Government Printing Office, April 1986). National data for 1999–2006 were downloaded from U.S. Environmental Protection Agency, AIRData, http://www.epa.gov/air/data/geosel.html (accessed March 21, 2007).

NOTES: The horizontal line marks the 15 µg/m³ federal standard. The 1981 point represents an average of 1979–1983, based on dichotomous sampler data from the Inhalable Particulate Network (IPN) from each of eighty-six counties, mostly in metropolitan areas. Data for 2000–6 are for the same eighty-seven counties, but with data collected by Federal Reference Method (FRM) samplers that are now used to determine compliance with federal PM₂.₅ standards. The 1981 IPN values include a correction for the low bias of dichotomous samplers relative to the FRM (see text and notes for explanation). For comparison, national-average FRM measurements for 269 monitoring sites with continuous data for 1999–2006 are included as well. These data show that the IPN locations do a good job of representing national trends.

of sites exceeding the annual standard declined from 30 percent to 13 percent.[7] Once again, areas with the highest levels achieved the greatest improvements. The appendix to this chapter on page 209 provides data on state-by-state trends for this period.

Nearly the entire country reduced $PM_{2.5}$ levels between 1999 and 2006, with declines occurring at 93 percent of the monitoring sites with continuous data. Of the locations that exceeded the 15 µg/m^3 national standard in 1999, 99 percent reduced their $PM_{2.5}$ levels between 1999 and 2006.

Trend data in a few areas of the country that monitored $PM_{2.5}$ during the 1980s and '90s also demonstrate ongoing improvements. For example, in Sand Mountain, Alabama, $PM_{2.5}$ declined 28 percent between 1982 and 1991.[8] In Washington, D.C., the reduction was 35 percent between 1993 and 2003, and in Chicago it was 29 percent between 1996 and 2003.[9] In California, $PM_{2.5}$ declined 50 percent between the late 1980s and 2004.[10]

EPA's 24-hour $PM_{2.5}$ standard requires that the ninety-eighth percentile of daily readings in each year average no more than 65 µg/m^3 over the most recent three-year period.[11] The ninety-eighth percentile is roughly the seventh-worst daily level measured in each year.[12] Virtually the entire nation complies with EPA's 24-hour $PM_{2.5}$ standard (all but one monitoring location as of the end of 2006). Recent data nevertheless show that daily levels also continue to decline. The national average of ninety-eighth percentile readings declined 17 percent from 1999 to 2006. IPN data also demonstrate long-term declines in peak $PM_{2.5}$ levels. Ninety-eighth percentile $PM_{2.5}$ levels declined an average of 45 percent between the early 1980s and 2006.

EPA has also collected data specifically on sulfate PM. Sulfate PM is formed in the atmosphere from gaseous SO_2 emissions, which come from burning coal for electricity. Virtually all sulfate PM falls into the $PM_{2.5}$ size range and contributes about 25–45 percent of $PM_{2.5}$ in the eastern half of the United States, where coal is a common electricity fuel, and about 10 to 20 percent in the western United States.[13] EPA's Clean Air Status and Trends Network (CASTNET) has monitored sulfate PM levels at a few dozen locations around the United States since the late 1980s.

Figure 3-2 on the following page displays the national-average sulfate trend for twenty-seven CASTNET sites that have continuous data from 1989 to 2006, and fifty-seven sites with continuous data from 1997 to 2006. As the chart shows, sites that were added during the mid-1990s tended to be in areas with relatively low sulfate levels. These are mainly in the western half of the country.

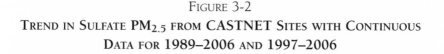

FIGURE 3-2
TREND IN SULFATE PM₂.₅ FROM CASTNET SITES WITH CONTINUOUS
DATA FOR 1989–2006 AND 1997–2006

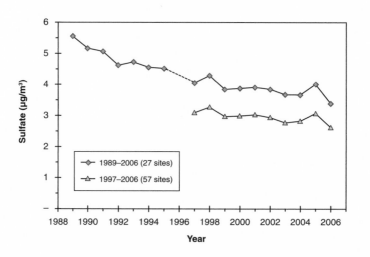

SOURCE: Data downloaded from EPA's CASTNET data Web site, www.epa.gov/castnet (accessed August 26, 2007).

NOTES: There is no data point for 1996. This is due to a budget impasse that resulted in the shutdown of many federal agencies for several weeks. CASTNET was one of the affected programs, and the 1996 CASTNET data were incomplete as a result. Since sulfate levels vary by season, this makes it impossible to calculate a valid annual average for the affected monitoring sites.

Average sulfate levels declined 34 percent from 1989 to 2006 and 12 percent from 1997 to 2006, based on a linear trend. Among the twenty-seven sites with data from 1989 onward, sulfate declined not only on average, but at every single location in the network. Areas with the highest levels achieved the largest declines.

We now turn to current PM₂.₅ levels across the United States. The results we present below are based on analysis of data from 791 monitoring sites with valid data for 2003–5. Figure 3-3 gives the range of annual PM₂.₅ levels in each state. The diamond marks the average level for all monitors in a given state, and the vertical line marks the range from worst to best monitoring site.

FIGURE 3-3

RANGE OF AVERAGE ANNUAL PM$_{2.5}$ LEVELS IN EACH STATE (PLUS THE
DISTRICT OF COLUMBIA), 2003–5

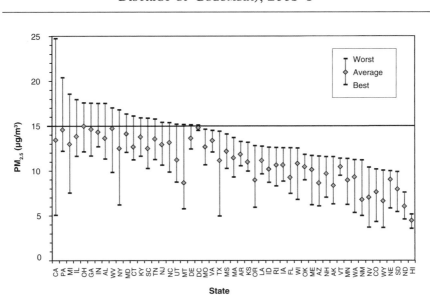

SOURCE: Analysis of site monitoring data downloaded from EPA's AIRData database, http://www.epa.gov/air/
data/geosel.html (accessed September 29, 2006).

NOTES: The horizontal line marks the 15 μg/m^3 federal standard. The graph was constructed by averaging
the three annual PM$_{2.5}$ readings for 2003–5 at each site and then calculating the average and range of these
values for each state.

Seventeen states (including the District of Columbia), or 33 percent,
have at least one location that violates the annual PM$_{2.5}$ standard. The average location complies with the standard in nearly all states, and most
violating sites are relatively close to the standard. California's worst
PM$_{2.5}$ were in a class by themselves until the last couple of years. However,
California has also reduced PM$_{2.5}$ more rapidly than the rest of the country,
so Pennsylvania now vies with California for the worst PM$_{2.5}$ levels.

Figure 3-4 also displays state-by-state PM$_{2.5}$ ranges, this time for
24-hour PM$_{2.5}$ levels. The federal standard is based on the ninety-
eighth percentile of 24-hour PM$_{2.5}$ readings each year, so we present the
average and range of ninety-eighth percentile readings for each state for

FIGURE 3-4
RANGE OF DAILY PM$_{2.5}$ LEVELS BY STATE
(PLUS THE DISTRICT OF COLUMBIA), 2003–5

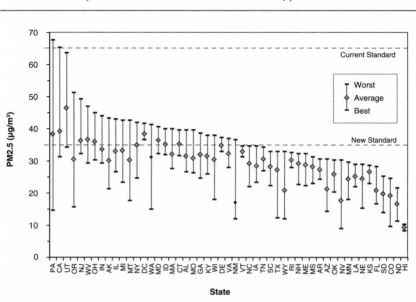

SOURCE: Analysis of site monitoring data downloaded from EPA's AIRData database, http://www.epa.gov/air/data/geosel.html (accessed September 29, 2006).

NOTES: The upper horizontal line marks the current federal standard of 65 µg/m^3. The lower horizontal line marks the proposed new standard of 35 µg/m^3. Levels for PM$_{2.5}$ are based on the ninety-eighth percentile of 24-hour values for each year (roughly the seventh-highest daily reading each year), which is how compliance with the federal standard is determined. The graph was constructed by averaging the three ninety-eighth percentile PM$_{2.5}$ readings for 2003–5 at each site and then calculating the average and range of these values for each state.

the 2003–5 period. Note that virtually the entire nation attains the current 24-hour standard of 65 µg/m^3, in most cases with plenty of room to spare. However, most states have areas that will violate the new standard of 35 µg/m^3 when EPA begins enforcing it in 2010.[14] Based on current PM$_{2.5}$ levels, twenty-nine of fifty-one states (including the District of Columbia) would have at least one monitoring location in violation of the new standard, compared with seventeen states under the current standard. The fraction of monitoring sites violating PM$_{2.5}$ standards would increase from 14 percent to 24 percent. The fraction of metropolitan areas in nonattainment would increase from 15 percent to 22 percent.

Note that Pennsylvania now has the worst 24-hour $PM_{2.5}$ site in the country. This site is in Liberty, Pennsylvania, in the Pittsburgh metro area. Liberty's $PM_{2.5}$ levels have been relatively stable over the last few years, while $PM_{2.5}$ has been dropping in California's high-$PM_{2.5}$ areas. Liberty's $PM_{2.5}$ levels are uniquely high for Pennsylvania, however. The next worst Pennsylvania location, a rural site in Allegheny County, averaged 52 $\mu g/m^3$ during 2003–5, or 25 percent lower than Liberty.

Logan, Utah, had uniquely high (for that area) peak 24-hour $PM_{2.5}$ levels in 2004 due to an unusually long period of stagnant air in January 2004.[15] Logan's high 24-hour $PM_{2.5}$ levels are likewise unique for Utah. The next worst Utah location, West Valley in the Salt Lake City metro area, had a ninety-eighth percentile $PM_{2.5}$ value of 48 $\mu g/m^3$ for 2003–5. Barring a similar bout of unusual meteorology, Utah will rank much lower for $PM_{2.5}$ in future years. Indeed, 2006 $PM_{2.5}$ data were becoming available as this book was going to press, and Logan's ninety-eighth percentile level for 2006 was only 29 $\mu g/m^3$.

Trends in PM_{10}

PM_{10} has been monitored since the late 1980s, although few sites have long-term data. Figure 3-5 on the following page displays the national trend in average annual PM_{10} levels. The heavy dotted line gives these levels for all sites that happened to be operating in a given year—a few hundred sites in the late 1980s, rising to about 1,500 in the mid-1990s, then dropping to about 1,000 today.

Despite the large number of sites operating at any given time, few have operated continuously for long periods. Thus, we have broken up the trends into two periods: 1988–96 (62 sites; 30 states) and 1993–2004 (156 sites; 33 states). All regions of the country are represented, and the period of overlap suggests that both groups of sites are representative of trends in PM_{10} levels across the country. Figure 3-5 also shows more recent trends and a wider array of locations by including data for 441 sites that operated from 1999 to 2004.

FIGURE 3-5

NATIONAL TREND IN AVERAGE ANNUAL PM_{10} LEVELS, 1987–2004

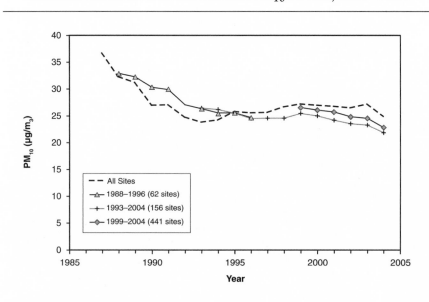

SOURCE: Analysis of site monitoring data downloaded from EPA's AIRData database, http://www.epa.gov/air/data/geosel.html (accessed January 15, 2006).

Based on the spliced trends for 1988–96 and 1993–2004, national-average PM_{10} levels dropped more than 30 percent from 1988 to 2004, and about 13 percent from 1999 to 2004, roughly mirroring the trend for $PM_{2.5}$.

Note that the trend based on all sites is different from that based on continuous sites. For all sites, PM_{10} declines through 1993, but then rises for a few years, leveling off after 1998. The data from continuous sites show that this apparent trend is not real. It turns out to be an artifact of the way the monitoring locations came into and out of existence during the last decade. As a result of the large reductions from the late 1980s to the mid-1990s, the vast majority of PM_{10} locations came into attainment, and many were retired. On the other hand, a small number of areas of the country that had persistently high PM_{10} levels tended to add monitoring sites to provide more detailed information on PM_{10} levels within the region. The addition

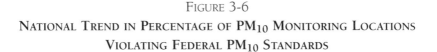

FIGURE 3-6

NATIONAL TREND IN PERCENTAGE OF PM$_{10}$ MONITORING LOCATIONS
VIOLATING FEDERAL PM$_{10}$ STANDARDS

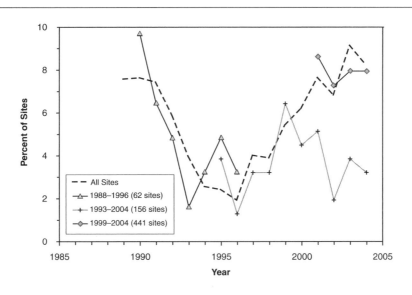

SOURCE: Analysis of site monitoring data downloaded from EPA's AIRData database, http://www.epa.gov/air/data/geosel.html (accessed January 15, 2006).

of sites in high-PM areas and their removal in low-PM areas creates a false appearance that PM$_{10}$ was not declining.

Figure 3-6 displays the trend in the percentage of monitoring sites that violate either the annual or 24-hour PM$_{10}$ standard. Because violation of a standard is based on three years of data, the trend begins in the third year of a given period (for example, 1990 for the 1988–96 trend). The continuous sites show that PM$_{10}$ violations dropped precipitously into the early 1990s and have since fluctuated between 2 and 6 percent. The 1999–2004 trend shows that the higher average violation rate is due to the recent addition of more monitors in high-PM$_{10}$ areas rather than a real increase in the extent of high PM$_{10}$ levels.

Violation of federal PM$_{10}$ standards is mainly localized in a few areas. California is home to the locations with the worst PM$_{10}$ in the country, with

Arizona, Texas, and Missouri having a few of the sites in the top twenty. As mentioned earlier, PM_{10} is more likely to be high in rural areas, largely due to windblown dust. For example, the highest PM_{10} areas of California (and the nation) are in rural Inyo and Mono counties.

Conclusion

In chapter 1 we showed that particulate levels declined throughout the twentieth century in those cities that have long-term data. Here we have shown that these improvements have continued during the last few decades. During the last twenty-five years, average PM levels have declined more than 40 percent, and about 15 percent since 1999. Back in the early 1980s, nearly the entire nation would have violated current $PM_{2.5}$ standards. But today, the violation rate is down to about 13 percent.

4

Why Air Pollution Will
Continue to Decline

The last three chapters showed that air pollution of all kinds has been declining for as long as we've been measuring it. Here we show that most remaining pollution emissions will be eliminated during the next twenty years *even if no new policy steps are taken*. Turnover in the auto and truck fleet from older vehicles to new, near-zero-emissions vehicles will eliminate more than 80 percent of remaining motor vehicle emissions, even after accounting for predicted growth in total miles of driving. Existing regulations require similar reductions for off-road diesel vehicles. Declining caps will eliminate more than 70 percent of remaining power plant emissions. This chapter reviews these and other underreported examples of large pollution reductions that will steadily occur over the next two decades.

Overview of National Pollution Emission Trends

America has dramatically reduced ambient air pollution levels by reducing emissions from all sources of pollution, both major and minor. EPA estimates that total emissions of the six "criteria" pollutants, plus VOCs, declined 53 percent between 1970 and 2005.[1] Lead was virtually eliminated as an air pollutant after it was fully eliminated from gasoline in the late 1980s. The vast majority of particulate matter emissions were also eliminated, with PM_{10} declining 84 percent since 1970. Emissions of VOC, SO_2, and CO all declined more than 50 percent, while NOx declined 30 percent since 1970.

Table 4-1 on the following page gives EPA's estimates of the percentage change in emissions of each pollutant for 1970–2005. To see what has

51

TABLE 4-1

EPA ESTIMATES OF EMISSIONS TRENDS FOR "CRITERIA" POLLUTANTS

Pollutant	Entire Period 1970–2005	Two Separate Periods 1970–1995	1995–2005
VOC	−53%	−36%	−26%
NOx	−29%	−8%	−23%
SO$_2$	−52%	−40%	−19%
CO	−55%	−39%	−26%
Lead	−99%	−98%	−25%
PM$_{10}$	−84%	−75%	−35%
PM$_{2.5}$	NA	NA	−9%

SOURCE: U.S. Environmental Protection Agency, *Air Emissions Trends—Continued Progress through 2005*, http://www.epa.gov/airtrends/econ-emissions.html (accessed November 27, 2006).

NOTES: Percentage reductions are calculated using emissions in the beginning year of each time period as the baseline (that is, 1970 or 1995). PM$_{10}$ and PM$_{2.5}$ emission trends include only *direct* emissions of particulates. Recall that much PM is "secondary" PM formed in the atmosphere from gaseous VOC, SO$_2$, and NOx emissions. Most of this secondary PM is PM$_{2.5}$. VOCs are not a criteria pollutant but are included here because they help form ozone and particulate matter, and most HAPs are also VOCs. NA = no estimates available.

happened more recently, the table also breaks the trend into two periods: 1970–95 and the last decade, 1995–2005. The percentage change is based on using the respective starting year, 1970 or 1995, as the baseline for determining subsequent percentage changes.

Note that while almost all lead emissions and most PM$_{10}$ had already been eliminated by the early 1990s, most NOx reductions occurred after 1995. Substantial improvements in VOC, SO$_2$, and CO occurred both before and after 1995. PM emissions declined relatively steadily during the last decade as well, but at a lower rate than for other pollutants.

The precise figures in table 4-1 create an unwarranted appearance of certainty. Keep in mind that they are estimates with significant uncertainties and some known biases. While it is clear that the United States has achieved large reductions in emissions of all pollutants, the actual reductions over time may in some cases be several percentage points greater or smaller than reported in table 4-1. We take up these accuracy issues in more detail later in this chapter.

Requirements that have already been adopted ensure that pollution will continue to decline. EPA reduced allowable emissions from new automobiles in 1994, 2001, and 2004. Most of the benefits of these standards have yet to be realized, because it takes fifteen to twenty years for the automobile fleet to turn over. On-road measurements show that emissions of the average automobile—including SUVs and pickup trucks—are dropping about 10 percent per year, while driving is increasing less than 2 percent per year, resulting in large net declines in total automobile emissions. As a result, total automobile emissions are on the decline even in rapidly growing metropolitan areas. The average automobile will get about 90 percent cleaner over the next two decades.

Likewise, EPA tightened standards for new heavy-duty diesel trucks in 1994, 1998, and 2003. A new round of standards will be implemented in 2007, requiring a 90 percent reduction in NOx and soot emissions when compared with the 2003 requirements. EPA also began tightening standards for new off-road diesel vehicles in 1996. The fourth round of off-road standards will come into force in 2010, reducing NOx and soot emissions more than 90 percent below current requirements.

Industrial emissions will also continue to decline. EPA requirements for power plants reduced SO_2 by 33 percent and NOx by 37 percent between 1990 and 2004. Power plant NOx emissions during the May–September "ozone-season" have declined 60 percent since 1998. EPA's Clean Air Interstate Rule will reduce SO_2 an additional 70 percent and NOx an additional 50 percent below recent levels during the next two decades.

Emissions of dozens of other pollutants—the so-called "hazardous air pollutants" (HAPs)—have also been declining. For example, emissions of mercury have declined 80–90 percent since the early 1980s, mainly due to reductions in emissions from waste incineration and ore processing. Coal-fired power plants are the only remaining substantial source of mercury, and they must reduce their mercury emissions by 70 percent by 2018 under EPA's Clean Air Mercury Rule.

Dozens of different VOC HAPs are emitted by motor vehicles, factories, small businesses, and consumer products, and all have declined as a result of efforts to control ozone and PM. In addition, under Title III of the Clean Air Act Amendments of 1990, EPA has required dozens of industries to install "maximum achievable control technology" (MACT) to reduce about

180 HAPs specifically listed in the Clean Air Act. These MACT requirements have eliminated most industrial emissions of a wide range of HAPs, including many specific VOCs (such as benzene), particulates (such as coke oven emissions), metals (such as chromium and mercury), and inorganic compounds (such as hydrogen fluoride). Several new MACT requirements come into effect over the next few years, assuring additional HAP reductions.

Compared to peak emissions during the 1960s and '70s, the vast majority of air pollution has already been eliminated. Already-adopted requirements will eliminate most of what remains during the next two decades.

Emissions Sources

EPA estimates that motor vehicles contribute about 54 percent of all national NOx emissions, with diesel trucks, automobiles, and off-road vehicles each contributing roughly a third of the mobile-source total. Electric utilities and industrial boilers add another 22 and 14 percent, respectively.[2] For SO_2, electric utilities account for 69 percent and industrial boilers 14 percent of total emissions.

EPA estimates motor vehicles contribute 45 percent of total VOC emissions, with the vast majority coming from gasoline vehicles rather than diesels. Industrial, commercial, and domestic solvent use (for example, in paints and cleaners) contributes another 30 percent to the estimated inventory, while VOC storage and transport (of gasoline and other petroleum products, among others) adds 8 percent. Motor vehicles are the overwhelming source of CO emissions, accounting for an estimated 91 percent.[3] Thus, for each pollutant, a small number of major source categories accounts for the vast majority of total emissions.

Although EPA provides precise estimates of pollutant emissions, most of these inventories have large uncertainties, and, in some cases, large known errors. The sulfur dioxide inventory is probably the most accurate, because most emissions come from a well-defined group of smokestacks at power plants and industrial facilities and are often measured directly with continuous monitors.

On the other hand, motor-vehicle emissions inventories are notoriously inaccurate and in some cases contain substantial known biases.[4] For example, although EPA's official estimate is that motor vehicles contribute 45 percent of all VOCs, field studies indicate that mobile sources, mainly gasoline vehicles, contribute 50–80 percent, depending on the metropolitan area.[5]

The NOx inventory might also have accuracy problems, but NOx emissions source contributions are more difficult than VOCs to assess with field studies. The reason is that VOCs include hundreds of different compounds, and each VOC source—automobiles, paints, refineries—has its own "signature" based on the particular compounds it emits. This allows researchers to estimate the contributions of various sources based on the amounts of each VOC they measure in the air. However, NOx is NOx, and it looks the same no matter where it comes from. While a number of studies have assessed relative NOx emissions from various mobile sources based on on-road measurements and fuel sales data, the accuracy of the overall NOx inventory is not known.

The contribution of various source categories to total emissions varies a great deal from place to place. Although EPA estimates that mobile sources contribute about 56 percent of NOx on a nationwide basis, their estimated contribution is generally higher in urban areas, typically ranging from about 60 to 90 percent.[6] For example, the air pollution regulatory agency for the Los Angeles–San Bernardino area estimates that motor vehicles account for 90 percent of regional NOx emissions.[7]

Emissions can also vary a great deal by region. For example, power plants contribute a larger fraction of NOx emissions in the eastern half of the United States, where coal is a major electricity source, than they do in the West, where coal is less common. For the same reason, most emissions of sulfur dioxide also occur in the eastern half of the United States.

Continuing Declines in Motor Vehicle Emissions

EPA has progressively tightened automobile and diesel-truck emissions standards during the last few decades. Figures 4-1 and 4-2 on pages 57 and 58 show the progression of automobile standards since the 1960s. Markings along the bottom of each figure give EPA's designation for each standard's regime and the time period during which it was in effect. As the figures show, automobiles produced today are about 99 percent cleaner than models produced before the 1970s. The same will be true of heavy-duty diesel trucks starting in 2007 and of off-road diesel vehicles starting in 2010. Figure 4-3 on page 59 shows the progression of NOx and PM standards for heavy-duty diesel trucks.[8] While diesel vehicles are not major sources of VOCs or CO, EPA's diesel standards also require similarly large reductions in these pollutants.

The improvement in automobile emissions over time is even greater than suggested by figures 4-1 and 4-2. Before Tier 2, the "large SUV/pickup" category in the charts included vehicles up to a gross vehicle weight rating (GVWR) of 5,750 pounds.[9] Under Tier 2, this category was expanded to include pickups up to a GVWR of 8,500 pounds, which would include pickups as large as, say, a Ford F-150, and SUVs up to a GVWR of 10,000 pounds, which includes the Hummer. In addition, while the Tier 1 standards applied up to 100,000 miles, the Tier 2 standards apply up to 120,000 miles and allow less emissions deterioration than the Tier 1 standards.

Figures 4-1 and 4-2 refer only to tailpipe standards. But without emissions controls, automobiles also emit large amounts of non-tailpipe VOC. Tier 1 and Tier 2 standards progressively eliminated these as well, with the level of control being at least as stringent as for tailpipe emissions.[10] The technological transition from carburetion to fuel injection during the late 1980s and early 1990s also eliminated a great deal of non-tailpipe VOC emissions.

The Tier 2 standards also require that the sulfur content of gasoline be reduced by about 90 percent. Since sulfur reduces the efficiency of catalytic converters and other components, lower sulfur content will likely reduce emissions from pre-Tier 2 automobiles as well.

Although new automobiles have near-zero emissions, many of the benefits of progressively tighter automobile standards are yet to be realized. This is because, as mentioned above, it takes fifteen to twenty years for the automobile fleet to turn over. Thus, many vehicles built to laxer standards are still on the road. Emissions of the average automobile will continue to decline as these earlier models age and head for the scrap heap. The same is true for the diesel vehicle fleet.

Real-world emissions data demonstrate the effect of this progression toward tighter standards for new vehicles. For example, figure 4-4 on

FIGURE 4-1

FEDERAL VOC EMISSIONS STANDARDS FOR CARS AND LIGHT TRUCKS

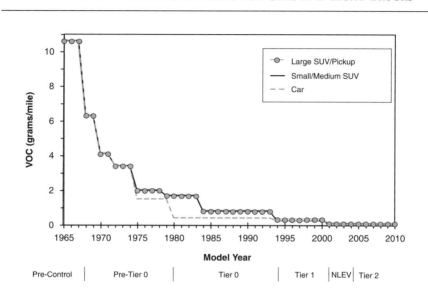

SOURCES: D. Bearden, *Air Quality and Vehicle Emission Standards: An Overview of the National Low Emission Vehicle Program and Related Issues* (Washington, D.C.: Congressional Research Service, January 4, 1999), http://www.ncseonline.org/nle/crsreports/air/air-23a.cfm (accessed June 19, 2005); J. G. Calvert, J. B. Heywood, R. F. Sawyer, and J. H. Seinfeld, "Achieving Acceptable Air Quality: Some Reflections on Controlling Vehicle Emissions," *Science* 261 (1993):37–45; S. C. Davis and S. W. Siegel, *Transportation Energy Data Book: Edition 22* (Oak Ridge, Tenn.: Oak Ridge National Laboratory, September 2002), www.cta.ornl.gov/cta/data/Download22.html (accessed May 3, 2003); Environmental Protection Agency, *Federal and California Exhaust and Evaporative Emission Standards for Light-Duty Vehicles and Light-Duty Trucks*, February 2000, http://www.epa.gov/otaq/cert/veh-cert/b00001.pdf (accessed June 19, 2005).

NOTES: Standards shown here apply up to 50,000 miles. Tier 1 added standards that apply between 50,000 and 100,000 miles, while Tier 2 added standards that apply between 50,000 and 120,000 miles. Tier 1 phased in during model years 1994–96. Tier 2 phased in during model years 2004–7. Designations along the bottom of the chart refer to the names EPA uses to refer to each set of standards. The NLEV, or National Low-Emission Vehicle program, was implemented nationwide in 2001. However, nine northeastern states implemented NLEV in 1999.

page 61 displays the trend for California automobiles (cars, SUVs, and pickup trucks) as measured in a highway tunnel in the San Francisco Bay Area from 1994 to 2001. Note the large declines in VOC, NOx, and CO emissions in just this seven-year period. Automobile benzene

FIGURE 4-2

FEDERAL NOX EMISSIONS STANDARDS FOR CARS AND LIGHT TRUCKS

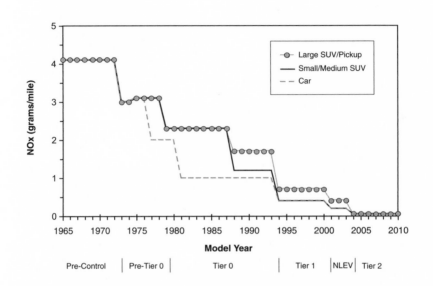

SOURCES: D. Bearden, *Air Quality and Vehicle Emission Standards: An Overview of the National Low Emission Vehicle Program and Related Issues* (Washington, D.C.: Congressional Research Service, January 4, 1999), http://www.ncseonline.org/nle/crsreports/air/air-23a.cfm (accessed June 19, 2005); J. G. Calvert, J. B. Heywood, R. F. Sawyer, and J. H. Seinfeld, "Achieving Acceptable Air Quality: Some Reflections on Controlling Vehicle Emissions, *Science* 261 (1993): 37–45; S. C. Davis and S. W. Siegel, *Transportation Energy Data Book: Edition 22* (Oak Ridge, Tenn.: Oak Ridge National Laboratory, September 2002), www.cta.ornl.gov/cta/data/Download22.html (accessed May 3, 2003); Environmental Protection Agency, *Federal and California Exhaust and Evaporative Emission Standards for Light-Duty Vehicles and Light-Duty Trucks*, February 2000, http://www.epa.gov/otaq/cert/veh-cert/b00001.pdf (accessed June 19, 2005)..

NOTES: Standards shown here apply up to 50,000 miles. Tier 1 added standards that apply between 50,000 and 100,000 miles, while Tier 2 added standards that apply between 50,000 and 120,000 miles. Tier 1 phased in during model years 1994–96. Tier 2 phased in during model years 2004–7. Designations along the bottom of the chart refer to the names EPA uses to refer to each set of standards. The NLEV, or National Low-Emission Vehicle program, was implemented nationwide in 2001. However, nine northeastern states implemented NLEV in 1999.

emissions were also measured, and they declined 80 percent over the same period.

Data from other areas and using other measurement techniques give similar results.[11] Table 4-2 on page 62 summarizes the annual percentage

FIGURE 4-3
FEDERAL NOx AND SOOT STANDARDS FOR HEAVY-DUTY DIESEL TRUCKS

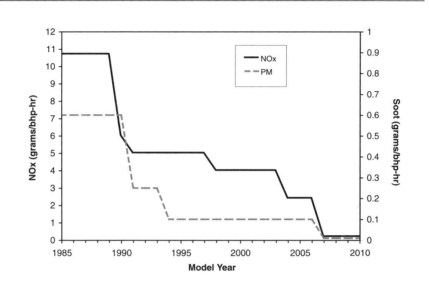

SOURCE: U.S. Environmental Protection Agency, *Health Assessment Document for Diesel Engine Exhaust* (Washington, D.C.: U.S. Environmental Protection Agency, May 2002).

reduction in average vehicle emissions based on data collected in several cities. VOC emissions have been declining 11–14 percent per year, NOx 3–9 percent per year, and CO 8–15 percent per year. An even longer-term trend is available for CO emissions from automobiles, based on data collected in Pennsylvania tunnels from 1973 to 1999. As shown in figure 4-5 on page 63, average CO emissions from automobiles declined 89 percent during this twenty-six-year period.

Average vehicle emissions rates are thus declining on the order of about 10 percent per year. At the same time, the total amount of driving continues to increase at a rate of nearly 2 percent per year, due to a combination of population growth and increasing driving per capita.[12] Will increases in driving offset the benefits of cleaner cars and trucks? It is not even close. Vehicles are getting cleaner much more rapidly than driving is increasing. Even after accounting for increasing travel,

Methods for Determining Real-World
Motor-Vehicle Emissions Trends

Real-world trends in motor-vehicle emissions can be estimated in three ways:

- *On-road remote sensing measurements.* A remote sensing device (RSD) is a technology that allows researchers to measure vehicles' emissions as they drive on the road. The remote sensor shoots a beam of light across the road, and the amount of pollution in a car's exhaust stream is determined based on how much of the light is absorbed as it passes through the exhaust plume. A single remote sensor can measure thousands of cars per day on a busy road. For several years, scientists from the University of Denver have been going back to the same locations in a few different cities, measuring emissions of tens of thousands of automobiles each year. RSD data allow researchers to look at trends in vehicle emissions by age and model year.

- *Highway tunnel studies.* In a tunnel study, researchers compare pollution levels in air entering and leaving a highway tunnel. The difference represents the emissions of the vehicles traveling through the tunnel. By going to the same tunnel year after year, scientists can assess trends over time in the average emissions of the vehicle fleet traveling in a given area.

- *Vehicle emissions inspection and maintenance (I/M) programs.* Each year, I/M programs measure large samples of vehicles of many model years, allowing researchers to see how average emissions of vehicles change as they age, and whether emissions performance has improved with successive model years.

Each of these methods has advantages and disadvantages. But taken together they give us a reasonable picture of trends in the emissions performance of motor vehicles.

total emissions have still been declining at a net rate of about 8 percent per year.

Figure 4-6 on page 64 illustrates this for California, based on the tunnel data discussed above. Between 1994 and 2001, the average vehicle's

FIGURE 4-4

TREND IN AUTOMOBILE EMISSIONS AS MEASURED IN A SAN FRANCISCO
BAY AREA TUNNEL, 1994–2001

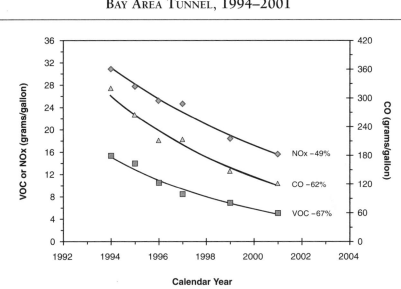

SOURCE: A. J. Kean, R. F. Sawyer, R. A. Harley, and G. R. Kendall, *Trends in Exhaust Emissions from In-Use California Light-Duty Vehicles, 1994–2001* (Warrendale, Penn.: Society of Automotive Engineers, 2002).

NOTES: Emissions are reported as grams of pollutant emitted per gallon of fuel burned. The curves drawn through the data points are exponential curves determined by a mathematical technique called least-squares fitting. The assumption of exponential decline seems more plausible than linear decline, because linear decline would result in zero emissions within a few years, which is unrealistic.

emissions rate (in grams of pollutant emitted per gallon of fuel burned) declined 67 percent for VOC, 49 percent for NOx, 62 percent for CO, and 80 percent for benzene. During the same period, gasoline use rose 13 percent in California.[13] The increase in gasoline consumption represents the combined effects of population growth, increases in per-capita driving, and lower fleet-average fuel economy due to the increasing popularity of SUVs and pickup trucks.

Combining the effects of lower vehicle emissions rates and increased gasoline use, Figure 4-6 shows that *total* vehicle emissions still declined 77 percent for benzene, 63 percent for VOC, 57 percent for CO, and 42 percent for NOx.[14] The chart also shows the growth in SUVs/pickups as

TABLE 4-2
ANNUAL PERCENTAGE REDUCTION IN AVERAGE VEHICLE EMISSIONS
RATES FOR VARIOUS TIME PERIODS, LOCATIONS, AND DATA SOURCES

Location	Time Period	VOC	NOx	CO	Number of Years Measured
San Francisco Bay Area Tunnel	1994–2001	14.9%	9.4%	12.8%	6
Chicago Remote Sensing	1997–2000	13.3%	8.5%	14.9%	4[a]
Denver Remote Sensing	1996–2003	NA	3.4%	8.3%	6[b]
Denver Emissions Inspections	1996–2002	13.2%	4.7%	10.1%	7
Phoenix Emissions Inspections	1995–1999	11.5%	5.3%	10.3%	5
Pennsylvania	1973–1999	NA	NA	8.6%	6

SOURCES: A. W. Gertler, M. Abu-Allaban, W. Coulombe, et al., "Measurements of Mobile Source Particulate Emissions in a Highway Tunnel," *International Journal of Vehicle Design* 27 (2002): 86–93; A. J. Kean, R. F. Sawyer, R. A. Harley, and G. R. Kendall, *Trends in Exhaust Emissions from In-Use California Light-Duty Vehicles, 1994–2001* (Warrendale, Penn.: Society of Automotive Engineers, 2002); T. Kirchstetter, A. Strawa, G. Hallar, et al., "Characterization of Particle and Gas Phase Pollutant Emissions from Heavy- and Light-Duty Vehicles in a California Roadway Tunnel" (paper, American Geophysical Union Fall Meeting, San Francisco, December 13–17, 2004); W. R. Pierson, "Automotive CO Emission Trends Derived from Measurements in Highway Tunnels," *Journal of the Air & Waste Management Association* 45 (1995): 831–32; J. Schwartz, *No Way Back: Why Air Pollution Will Continue to Decline* (Washington, D.C.: American Enterprise Institute, 2003), http://www.aei.org/docLib/20030804_4.pdf (accessed August 27, 2006.
NOTES:
a. Three years for VOC, spanning 1997–2000.
b. Four years for NOx, spanning 1999–2003.

a fraction of the vehicle fleet, demonstrating that large net declines in total vehicle emissions are occurring even as SUVs become an ever larger fraction of the automobile fleet. This is because, during the 1990s, a new SUV's emissions weren't that much greater than a new car's emissions. Data from vehicle emissions inspection programs demonstrate this. Figure 4-7 on page 65 compares emissions of cars and SUVs/pickups using inspection data from Denver. Note that SUVs have had the same VOC emissions as cars since about the 1996 model year, and the same NOx emissions in the 2001 model year, the most recent data publicly available. Inspection data from Phoenix and on-road data from California give similar results.[15]

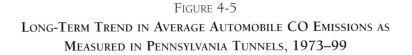

FIGURE 4-5

LONG-TERM TREND IN AVERAGE AUTOMOBILE CO EMISSIONS AS
MEASURED IN PENNSYLVANIA TUNNELS, 1973–99

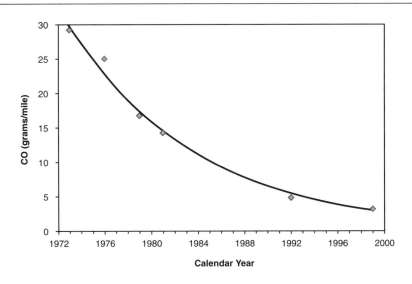

SOURCES: A. W. Gertler, J. A. Gillies, W. R. Pierson, et al., "Real-World Particulate Matter and Gaseous Emissions from Motor Vehicles in a Highway Tunnel," *Research Report/Health Effects Institute* (2002): 5–92; W. R. Pierson, "Automotive CO Emission Trends Derived from Measurements in Highway Tunnels," *Journal of the Air & Waste Management Association* 45 (1995): 831–32.

NOTES: Emissions are reported as grams of pollutant emitted per mile of travel. The curve drawn through the data points is an exponential curve determined by a mathematical technique called least-squares fitting. The assumption of exponential decline seems more plausible than linear decline, because linear decline would result in zero emissions within a few years, which is unrealistic.

Even in the early 1990s, before SUVs caught up with the VOC emissions of cars, turning over the fleet to new SUVs reduced emissions. At worst, the increasing popularity of SUVs slowed down the rate at which emissions declined, but it never caused them to increase.

SUVs have had about the same emissions as cars for the last few model years. Furthermore, EPA's Tier 2 and California's LEV II (for "Low Emission Vehicle II") requirements, both of which began phasing in with the 2004 model year, require SUVs and pickup trucks to meet the same emissions standards as cars.[16] Thus, for the last few years, whether Americans

FIGURE 4-6

CHANGE IN TOTAL EMISSIONS OF CALIFORNIA AUTOMOBILES, 1994–2001,
AFTER ACCOUNTING FOR GROWTH IN DRIVING AND SUV/PICKUP POPULARITY

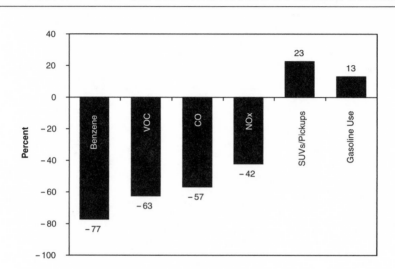

SOURCE: A. J. Kean, R. F. Sawyer, R. A. Harley, and G. R. Kendall, *Trends in Exhaust Emissions from In-Use California Light-Duty Vehicles, 1994-2001* (Warrendale, Penn.: Society of Automotive Engineers, 2002)

purchased SUVs, cars, or pickup trucks made no difference for air quality, and it will continue to make no difference.

Activists and regulators have been claiming that the growth in popularity of SUVs is offsetting the gains from cleaner vehicles. For example, a 2003 *USA Today* story asserted that Americans are driving "vehicles that give off more pollution than the cars they drove in the '80s."[17] A Sierra Club report, *Clearing the Air with Transit Spending*, claimed that gains from cleaner vehicles are being "canceled out" by suburbanization and SUVs.[18] The president of the antisuburb/antiautomobile activist group Smart Growth America claimed, "Sprawl and higher-emitting SUVs are proliferating faster than technological fixes can keep up."[19] Actual data on real-world vehicle emissions show these claims are the polar opposite of reality.

Long-term emissions trend data for diesel trucks show that diesel soot emissions, like emissions from SUVs, have also declined a great deal.

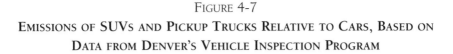

FIGURE 4-7

EMISSIONS OF SUVS AND PICKUP TRUCKS RELATIVE TO CARS, BASED ON
DATA FROM DENVER'S VEHICLE INSPECTION PROGRAM

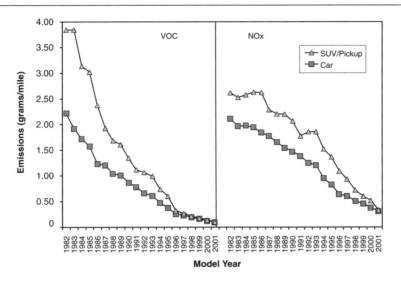

SOURCE: Averages of emissions by calendar year for each vehicle type and model year provided by Tom Wenzel, Lawrence Berkeley National Laboratory.

NOTES: Data are from the Denver vehicle emissions inspection program collected during June-August in calendar years 1996–2002.

Figure 4-8 on page 66 shows diesel-truck data collected in Pennsylvania tunnels.[20] The average emissions rate declined 83 percent from 1973 to 1999. Data from a tunnel in the San Francisco Bay Area indicate that the average rate declined at least 50 percent between 1997 and 2004.[21]

NOx emissions from heavy-duty diesel trucks (HDDT) is the only case in which the evidence suggests little or no change during the 1990s. Data collected in the Tuscarora Tunnel in Pennsylvania in 1992 and 1999 suggest per-mile HDDT NOx emissions may have risen.[22] On the other hand, estimates based on California data suggest modest declines in NOx emissions, while laboratory test data of in-use engines suggest they have remained roughly constant.[23] Overall, there was probably little change during the 1990s.

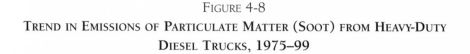

FIGURE 4-8

TREND IN EMISSIONS OF PARTICULATE MATTER (SOOT) FROM HEAVY-DUTY
DIESEL TRUCKS, 1975–99

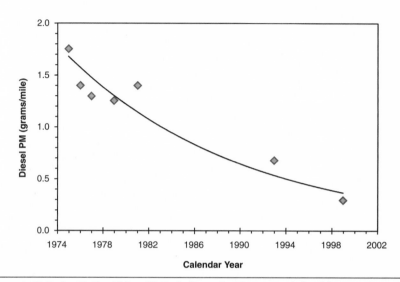

SOURCES: A. W. Gertler, M. Abu-Allaban, W. Coulombe, et al., "Measurements of Mobile Source Particulate Emissions in a Highway Tunnel," *International Journal of Vehicle Design* 27 (2002): 86–93; A. W. Gertler, J. A. Gilles, W. R. Pierson, et al., "Emissions from Diesel and Gasoline Engines Measured in Highway Tunnels," *Health Effects Institute*, January 2002, www.healtheffects.org/Pubs/GertGros.pdf (accessed May 14, 2004).

EPA tightened HDDT NOx standards in 1988 and 1991, so one might have expected to see declines during the 1990s.[24] However, engine manufacturers programmed their engines to maximize fuel economy during steady-state operation (i.e., during freeway driving). This increased NOx emissions as a side effect (but lowered particulate matter emissions, also as a side effect). This probably accounts for the lack of NOx reductions from diesel trucks. Although programming for higher fuel economy met the letter of EPA's engine-certification requirements, EPA ruled the practice illegal and, under a settlement agreement, the engine manufacturers agreed to reprogram affected engines when they came in for scheduled maintenance.[25]

The reprogramming issue applies mainly to 1993 through 1998 engine models, and the emission-control software for these engines can be updated when the vehicles come in for an engine rebuild, presumably bringing their NOx emissions down.[26] EPA tightened HDDT NOx standards with the 1998 model year (with a cap 20 percent below the 1991 standard), and again in 2003 (with a cap 50 percent below the 1991 standard).[27] Between the tougher standards for new trucks and the enforcement action for existing trucks, it seems reasonable to expect that HDDT NOx emissions have declined during the last few years. Recent measurements in California suggest that HDDT NOx emissions have indeed been dropping during the last few years.[28]

Although real-world trend data are also lacking for off-road diesel equipment, EPA estimates that from 1990 to 2002, total NOx from off-road diesels increased 10 percent (with total soot emissions decreasing 15 percent), peaking in 1997 and remaining flat since.[29] This is consistent with expected declines due to tougher standards that have been progressively implemented since 1996; however, real-world data are not available to substantiate the predictions.

Continuing Declines in Industrial Emissions

After mobile sources, power plants and industrial facilities are the next-largest group of contributors to total air pollution emissions. For example, EPA estimates that in 2002 electric utilities accounted for about 22 percent of all NOx emissions and 69 percent of all SO_2.[30] Industrial boilers contributed an estimated 14 percent of both in 2003.[31] Most of these emissions occurred in the eastern half of the United States, where coal is a major electricity fuel.

Figures 4-9 and 4-10 on pages 68 and 69 show the long-term trend in NOx and SO_2 emissions from power plants and industrial boilers.[32] NOx from industrial boilers has declined by about a third since 1970. NOx from power plants rose from 1970 to 1984, dropped in 1985, and then didn't change much until the late 1990s. Since 1998 total power plant NOx emissions are down about 45 percent, while ozone-season (May–September) emissions are down 55 percent. These recent NOx reductions were required by the Clean Air Act's Title IV acid rain program and, more recently, by EPA's "NOx SIP Call" regulation.[33] The NOx SIP Call is intended to reduce NOx during May through September, the period of the year when ozone levels are highest.[34]

FIGURE 4-9

TREND IN NOx EMISSIONS FROM ELECTRIC UTILITIES AND INDUSTRIAL
BOILERS, 1970–2006

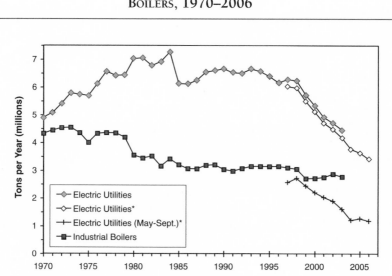

SOURCES: U.S. Environmental Protection Agency, "Clean Air Markets–Data and Maps," http://cfpub.epa.
gov/gdm/index.cfm?fuseaction=iss.isshome (accessed March 21, 2007); U.S. Environmental Protection
Agency, "1970–2002 Average Annual Emissions, All Criteria Pollutants in MS Excel," http://www.epa.
gov/ttn/chief/trends/ (accessed March 21, 2007); U.S. Environmental Protection Agency, "Nitrogen Oxides;
National Emission Totals," http://www.epa.gov/ttn/naaqs/ozone/ozonetech/airtrends/2005/pdfs/NOX
National.pdf (accessed March 21, 2007).

NOTES: The trend from 1970 to 2003 is from EPA's national emission inventories broken out by major
source categories. The year 2003 is the most recent period for which EPA has published consistent long-
term national inventory estimates. The emissions trend for 1997–2006 (marked with an asterisk in the leg-
end) is from EPA's "Clean Air Markets–Data and Maps" Web site, which allows visitors to download NOx
and SO$_2$ emissions estimates for electric utilities. As the graph shows, the estimate of total emissions from
electric utilities is slightly lower than the estimates from EPA's national inventory. This might be because
the universe of sources is slightly different (e.g., a few minor fuel types might not be included in the Clean
Air Markets inventory) or because of small differences in emissions estimation methods. We provide data
from both sources because the Clean Air Markets estimates go all the way through 2006. We also show the
Clean Air Markets emissions estimates for May–September because this corresponds to the "ozone season."

Industrial boiler SO$_2$ emissions have dropped by about half since 1970,
with most of the decline occurring during the early 1970s and the late 1990s.
For power plant SO$_2$, Title IV of the Clean Air Act created a "cap-and-trade"
program. Total emissions from coal-fired power plants were capped starting

FIGURE 4-10

TREND IN SO₂ EMISSIONS FROM ELECTRIC UTILITIES AND INDUSTRIAL
BOILERS, 1970–2006

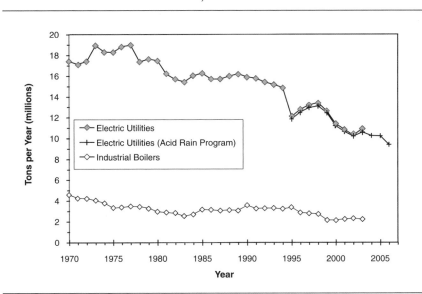

SOURCES: U.S. Environmental Protection Agency, "Clean Air Markets–Data and Maps," http://cfpub.
epa.gov/gdm/index.cfm?fuseaction=iss.isshome (accessed March 21, 2007); U.S. Environmental
Protection Agency, "1970–2002 Average Annual Emissions, All Criteria Pollutants in MS Excel," http://
www.epagov/ttn/chief/trends/ (accessed March 21, 2007); U.S. Environmental Protection Agency,
"Emissions Inventory Trends, 1970–2003, http://www.epa.gov/air/airtrends/aqtrnd04/econ-emissions.
html (accessed December 15, 2005; link no longer active).

NOTES: The trend from 1970 to 2003 is from EPA's national emission inventories broken out by major
source categories. The year 2003 is the most recent period for which EPA has published consistent national
inventory estimates. The emissions trend for 1995–2006 (the "Acid Rain Program" trend) is from EPA's
"Clean Air Markets–Data and Maps" Web site, which allows visitors to download NOx and SO₂ emissions
estimates for electric utilities. As the graph shows, the estimate of total emissions from electric utilities is
slightly lower than the estimates from EPA's national inventory. This might be because the universe of
sources is slightly different (e.g., a few minor fuel types might not be included in the Clean Air Markets
inventory) or because of small differences in emissions estimation methods. We provide data from both
sources because the Clean Air Markets estimates go all the way through 2006.

in 1995, and the cap declined each year thereafter. Power plants were
required either to reduce emissions each year to meet their portion of the
reductions necessary to stay under the cap, or to acquire "allowances" from
other facilities that had reduced beyond their requirements. The declining

cap is responsible for the large reductions in SO_2 emissions after 1994—which are, in turn, responsible for decreasing levels of SO_2 and sulfate PM in the air.

Note, however, that after a sharp drop in 1995, SO_2 emissions increased for a few years, before declining again after 1998. This was due to the Title IV program's "banking" provision. Facilities that reduced emissions below their cap in one year were allowed to "bank" some of those reductions for use in future years. Banked emissions allowances expire after a fixed number of years, ensuring continued overall declines.

When SO_2 emissions increased slightly in 2003, some environmental groups claimed this was the result of the Bush administration's "roll back" or "gutting" of Clean Air Act requirements.[35] In fact, it had nothing to do with Bush administration air policies, just as the 1995–98 SO_2 increase (and the decline thereafter) was not the result of any Clinton administration policy. Both trends were due to the banking provision of the SO_2 trading program enshrined in the Clean Air Act Amendments of 1990.

EPA has also drastically reduced VOC emissions from industry. For example, as part of its regulation of hazardous air pollutants, EPA in 1997 required an 80 percent VOC reduction from organic chemical manufacturing, which amounted to more than 1,000,000 tons per year.[36] Refineries had to reduce emissions by 60 percent, or about 280,000 tons per year, in 1998.[37] Those are just the largest industrial sources of VOC pollution. Several dozen other requirements set limits on emissions from a wide range of other industries.[38] EPA also adopted regulations requiring reduced VOCs in paints, coatings, and cleaning fluids—the major VOC sources in consumer products.[39]

Title III of the Clean Air Act Amendments of 1990 required EPA to issue National Emission Standards for Hazardous Air Pollutants (NESHAPs) for a wide range of HAPs. By EPA's estimation, hexavalent chromium and coke-oven emissions together accounted for most of the cancer risk imposed by industrial emissions. According to EPA's estimates for 1996, at the ninety-ninth-percentile exposure level—that is, in areas with the highest levels of coke-oven emissions and hexavalent chromium—people had a lifetime risk of about 1 in 66,000 and 1 in 14,000 (0.0015 percent and 0.007 percent), respectively, of developing cancer due to each of these pollutants.[40] However, EPA implemented an 80 percent reduction in coke-oven emissions starting in 1996 and a 99 percent reduction in hexavalent chromium emissions starting in 1998.[41] In other words, even the small worst-case risks that EPA estimated for 1996 have since been largely eliminated.

Other NESHAPs have eliminated most emissions from dry cleaning, waste incineration, polymer production, lead smelting, copper and aluminum production, printing, oil and natural gas production, pesticide production, cement manufacturing, and dozens of other more arcane industrial facilities and processes. As shown in chapter 1, ambient levels of HAPs have been declining since measurements began in the early 1990s—just as would be expected based on these and other emission-reduction requirements.

Mercury has been a particular focus of regulators and environmentalists during the last few years for two main reasons. First, coal-fired power plants are now the largest source of mercury air emissions in the United States. Second, mercury in fish is believed to come largely from air emissions, and health experts are concerned that people might suffer cognitive and neurological harm from eating fish with elevated levels of mercury.[42]

As with other pollutants, the United States has already eliminated most mercury air emissions. Total United States mercury emissions have been reduced by about 80–90 percent since the early 1980s, and by 70 percent since 1990.

Most mercury air emissions during the 1980s resulted from incineration of mercury-containing wastes. With the removal of most mercury from the waste stream in the late 1980s and strict pollution regulations implemented for waste incinerators during the 1990s, coal-fired power plants have become the largest source of human-caused U.S. mercury emissions. Coal naturally contains trace amounts of mercury, which is emitted into the air when the coal is burned to produce energy.

EPA estimates that of the 112 tons of mercury emitted by the United States in 2002, 45 percent came from coal-fired utility boilers.[43] By comparison, in its 1997 *Mercury Study Report to Congress*, EPA estimated that 158 tons of mercury were emitted in 1995.[44] That inventory did not include emissions from gold ores, which were included in EPA's 1999 inventory. Assuming gold-ore emissions were the same in previous years, total emissions would have been 170 tons in 1995.

Due to much higher emissions from waste incineration and the roasting and milling of mercury ores, emissions were substantially higher in

1989 and 1990 than in subsequent years. Researchers from the United States Geological Survey (USGS) estimate that compared with 1995, emissions from waste incineration were 47 tons per year higher in 1990 and 162 tons per year higher in 1989.[45] Likewise, total emissions from mercury ore milling and roasting were virtually zero in 1996, but 72 tons in 1990. The USGS was not able to determine how much of the mining emissions were air emissions versus emissions to water or land.

Assuming emissions from other sources were the same in 1989–90 as in 1995, the 1989 and 1990 inventories would be, respectively, 374 and 259 tons per year.[46] The large decreases from the late 1980s to the early 1990s were due to drastic reductions in the amount of mercury in the waste stream coming into waste incinerators and to strict regulatory limits on mercury emissions from waste incineration. In addition, mercury ore mining in the United States had essentially ceased by the mid-1990s.[47]

We were not able to locate national emissions estimates for years prior to 1989. However, other evidence suggests emissions were substantially higher prior to 1989. Total mercury consumption was about 75 percent higher in the late 1970s and early 1980s when compared with 1989.[48] A mercury emissions inventory trend compiled for Florida concluded that emissions were 3.7 times higher in 1980 than in 1990, with the declines coming mainly from reductions in mercury from waste incineration.[49]

Figure 4-11 displays estimated mercury emissions for 1989 through 2002, along with a rough guess at the range of the emissions in the early 1980s, based on overall U.S. mercury use and the Florida inventory. Figure 4-12 on page 74 shows the amount of mercury used in the U.S. economy from 1970 to 1997. Note the sharp reduction in mercury usage from 1988 to 1991, largely due to the removal of mercury from batteries and paints.

Existing Requirements Will Eliminate Most Remaining Pollution Emissions

Air pollution emissions of all kinds have been dropping, and current ambient pollution levels are only a fraction of the much higher levels of the past. With already-adopted requirements eliminating most remaining emissions over the next two decades, pollution will continue to decline.

FIGURE 4-11
TREND IN U.S. MERCURY EMISSIONS TO 2002

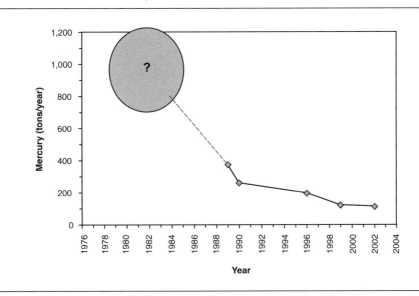

SOURCES: Environmental Protection Agency, *Mercury Study Report to Congress Volume II: An Inventory of Anthropogenic Mercury Emissions in the United States*, December 1997, http://www.epa.gov/ttn/oarpg/t3/reports/volume2.pdf (accessed November 27, 2006); J. D. Husar and R. B. Husar, "Trend of Anthropogenic Mercury Flow in Florida, 1930–2000" (paper, Mercury in the Environment: Assessing and Managing Multimedia Risks Meeting, American Chemical Society, Division of Environmental Chemistry, Orlando, Fla., April 7–11, 2002); U.S. Department of the Interior, Bureau of Mines, *The Materials Flow of Mercury in the United States*, by S. M. Jasinski, 1994, http://pubs.usgs.gov/usbmic/ic-9412/mercury.pdf (accessed November 27, 2006); U.S. Geological Survey, *The Materials Flow of Mercury in the Economies of the United States and the World*, by J. L. Sznopek and T. G. Goonan, 2000, http://pubs.usgs.gov/circ/2000/c1197/c1197.pdf (accessed November 27, 2006); U.S. Environmental Protection Agency, National Emission Inventory Documentation and Data (accessed March 15, 2007).

NOTE: Oval with question mark represents a likely range of national mercury emissions during the early 1980s, based on the amount of mercury in the waste stream and an emissions estimate for Florida.

Mobile sources—automobiles, diesel trucks, and off-road equipment—account for most nationwide NOx and VOC emissions. More than 80 percent of remaining mobile source emissions will be eliminated over the next twenty years or so, even after accounting for expected growth in motor-vehicle travel. Here's how we know this:

First, the rate of decline in ambient CO and NO_2 levels has been accelerating. For example, the annual decline in CO levels went from about

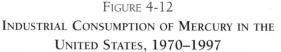

FIGURE 4-12
INDUSTRIAL CONSUMPTION OF MERCURY IN THE
UNITED STATES, 1970–1997

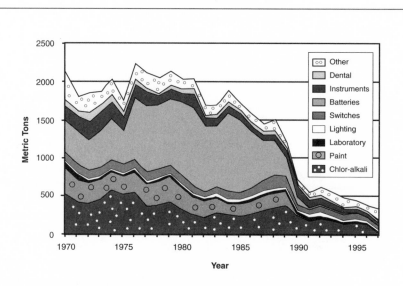

SOURCE: U.S. Geological Survey, *The Materials Flow of Mercury in the Economies of the United States and the World*, by J. L. Sznopek and T. G. Goonan, 2000, http://pubs.usgs.gov/circ/2000/c1197/c1197.pdf (accessed November 27, 2006).

4 percent during the mid-1980s to more than 7 percent today.[50] The rate of NO_2 decline went from 1 percent per year during the mid-1980s to nearly 3 percent per year today.[51]

CO and NO_2 are good indicators for what is happening with motor-vehicle emissions for three reasons: most emissions of these pollutants come from motor vehicles; they have been widely monitored for more than two decades; and most monitoring sites are located in urban or suburban areas where monitored levels would be directly affected by nearby motor-vehicle emissions.

Thus, accelerating decline in ambient levels of CO and NO_2 is evidence that the rate of decline in motor-vehicle emissions is also accelerating. This reasoning is particularly strong for CO, nearly all of which comes from motor vehicles.[52] And even for NO_2, motor vehicles make up most of the emissions. EPA estimates that motor vehicles account for 56 percent of NOx emissions on a nationwide basis and a much larger percentage of the total in most urbanized areas.[53]

Second, even if we had stopped with automobile standards that were in force in the late 1990s, emissions would still have dropped sharply, as vehicles built to earlier, less stringent standards were scrapped. We can see this by looking at data on the average performance of automobiles built in different model years, using data collected on the road via remote sensing or with tests performed in vehicle inspection programs.

When these data are disaggregated by model year, they show that with each successive year, the average automobile is starting out and staying cleaner than those built in previous years. That is, when compared at the same age, automobiles built more recently have lower emissions than automobiles built in previous years.

This is shown in figure 4-13 on page 76 with data from Denver's vehicle emissions inspection program, collected during calendar years 1996–2002. The graph shows average VOC emissions versus age for even model years 1982–2000. Cars and light trucks (SUVs and pickup trucks) are shown separately. Because each is measured at several ages, we can compare different model years at the same age. For example, in calendar years 1996–2002, the 1984 model year was measured at ages twelve through eighteen, while the 1986 model year was measured at ages ten through sixteen. Thus, both the 1984 and 1986 model years were measured at ages twelve through sixteen. Looking at the graph, we can see that from ages twelve through sixteen, 1986 model year SUVs were about 35 percent cleaner than 1984 model year SUVs.

A similar conclusion applies for all model years and ages: More recent models start out and stay cleaner than earlier models. Even if we had stopped with standards applicable to automobiles built in 2000, most automobile VOC emissions would still go away over the next fifteen to twenty years.

Third, we didn't stop tightening motor-vehicle emissions standards. EPA's National Low Emission Vehicle (NLEV) standards replaced the Tier 1 standards starting with the 2001 model year. EPA began implementing its Tier 2 standards with the 2004 model year, pushing emissions down further still. Compared with Tier 1, Tier 2 vehicles are required to be 77–95 percent cleaner over their lives.[54] The near-zero emissions levels required by the Tier 2 standards are shown in the first two figures of this chapter.

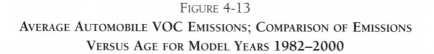

FIGURE 4-13

AVERAGE AUTOMOBILE VOC EMISSIONS; COMPARISON OF EMISSIONS
VERSUS AGE FOR MODEL YEARS 1982–2000

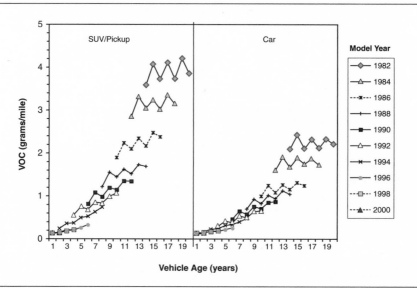

SOURCE: Denver vehicle emissions inspection program test results provided by Tom Wenzel, Lawrence Berkeley National Laboratory.

NOTES: Data from the Denver vehicle emissions inspection program collected during June–August in calendar years 1996–2002. The zigzag pattern results from the fact that cars from even-numbered model years are generally tested in odd calendar years only if they are sold. Cars that are sold tend to be in worse shape, and therefore tend to have higher emissions, than those that are not sold. Thus, for any even model year, the group of cars tested in odd calendar years will tend to have higher emissions than the group tested in even calendar years, creating the zigzag pattern.

In twenty years or so, when almost all automobiles on the road are Tier 2 automobiles, the average automobile will be about 90 percent cleaner than it is today. This can be seen by comparing current emissions with those of a fleet in which all automobiles are built to Tier 2 standards. According to EPA's MOBILE6 vehicle emissions model, the average automobile on the road in 2005 emitted about 1.8 grams/mile of VOC + NOx.[55] A fleet of vehicles meeting EPA's Tier 2 requirements and having the same age distribution as the current fleet would emit about 0.16 grams/mile of VOC + NOx, or 91 percent less than the estimate for the current fleet.[56]

FIGURE 4-14

LIKELY FUTURE TREND IN TOTAL AUTOMOBILE EMISSIONS IN A VERY HIGH GROWTH AREA, GIVEN RECENT IMPROVEMENTS IN AUTOMOBILE EMISSIONS PERFORMANCE AND DURABILITY AND NEW REGULATORY REQUIREMENTS

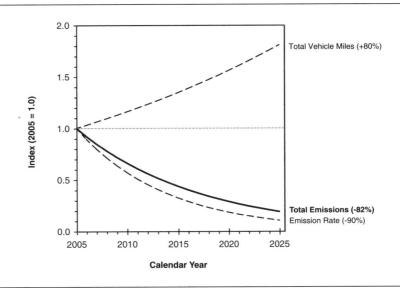

SOURCE: Author's calculations.

NOTES: The graph displays the future trend in total automobile emissions, assuming emissions of the average automobile decline 90 percent between 2005 and 2025 (a decline of a little over 10 percent per year) and total miles of driving increase 80 percent (an increase of about 3 percent per year). "Emission rate" is the emissions of the average automobile. "Total emissions" is the total emissions of the vehicle fleet after accounting for both declining emissions of the average automobile, and increasing total miles of driving. The chart indexes all three variables to a common value of 1.0 in 2005 and then shows the percentage change over time relative to 2005.

With reductions like these on the way, even large increases in total driving will have only a minor effect on future emissions. For example, if total miles driven increase 80 percent during the next twenty years, which might happen in a very high-growth metropolitan area, but per-mile emissions of the average vehicle decline 90 percent, then total emissions would still decline 82 percent.[57] Figure 4-14 displays the predicted future trend in total automobile emissions given the combined effects of progressively cleaner cars and increases in total driving. As the graph shows, automobiles will become so clean that even large increases in total driving will have little effect on future emissions.

Automobiles aren't the only mobile sources for which EPA has imposed tougher standards. Diesel trucks and off-road mobile sources have also been subject to progressively more stringent regulation. For example, EPA reduced NOx emissions limits for new diesel trucks by 40 percent in 2003, and requires an additional 90 percent reduction starting in 2007.[58] The 2007 regulation also requires a 90 percent reduction in soot emissions. These reductions are in addition to previous standards for diesel trucks implemented in the 1980s, 1990s, and, most recently, in 2003. MOBILE6 predicts that per-mile diesel truck NOx and PM emissions will decline about 90 percent between 2005 and 2025, which is in line with expectations, given the stringent requirements of the emissions standards.[59]

EPA's Tier 1 through Tier 3 rules for construction and farm equipment and other off-road diesel vehicles require a 70 percent NOx reduction and a 40 percent soot reduction in new diesels between 1996 and 2008.[60] EPA recently adopted its Tier 4 rule, which requires an additional 90 percent reduction in NOx and soot from off-road diesels, beginning in 2010.[61] Already-adopted EPA rules require large NOx reductions from marine vessels, locomotives, and off-road gasoline engines as well.[62]

Taken together, these requirements will eliminate the vast majority of remaining mobile-source emissions, even after accounting for growth.

Hazardous air pollutants from mobile sources will decline roughly in proportion to the declines in overall automobile and diesel emissions. The reason is that the main transportation-related HAPs are diesel soot and a few VOCs—for example, benzene and 1,3-butadiene. But we've already seen that current requirements will eliminate the vast majority of these pollutants as the fleet turns over to vehicles meeting the new emissions standards.

Although regulators and planners do not publicize it, they also predict large reductions in motor-vehicle emissions despite large predicted increases in total driving. These predictions are embodied in metropolitan areas' transportation conformity findings, which must be put in official reports that are updated every few years.[63]

Figure 4-15 summarizes these projections for five large metropolitan areas in various parts of the country. The graph displays the estimated percent change from 2005 to 2025 for total miles of driving, and total emissions of VOCs and NOx for all on-road vehicles, both gasoline and diesel.

FIGURE 4-15

GOVERNMENT PROJECTIONS OF THE CHANGE IN TOTAL VEHICLE MILES
TRAVELED AND TOTAL VEHICLE EMISSIONS BETWEEN 2005 AND 2025

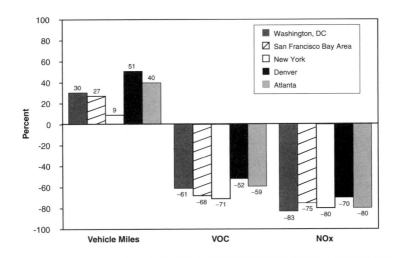

SOURCES: Metropolitan Transportation Commission, *Transportation Air Quality Conformity Analysis*, November 1, 2005, http://apps.mtc.ca.gov/meeting_packet_documents/agenda_574/tmp-3734_Draft_Conformity Analysis_2005TIP_05-16_final.doc (accessed December 7, 2006); Denver Regional Council of Governments, *2003 Amendments to the Metro Vision Plan and the Fiscally Constrained 2025 Interim Regional Transportation Plan*, August 20, 2003, http://www.drcog.org/documents/2025_RTP_Update_08-20-03.pdf (accessed December 7, 2006); Atlanta Regional, *Conformity Determination Report*, February 2006, http://www.atlantaregional.com/cps/rde/xbcr/SID-3F57FEE7-1C6BDB0C/arc/8hrOzone_13County.pdf (accessed December 7, 2006); National Capital Region Transportation Planning Board and Metropolitan Washington Council of Governments, *Air Quality Conformity Determination of the 2004 Constrained Long Range Plan and the FY-2005–2010 Transportation Improvement Program for the Washington Metropolitan Region*, November 17, 2004, http://www.mwcog.org/uploads/pub-documents/9VxbXA20041216154934.pdf (accessed December 7, 2006).

NOTES: Percentages are the projected change in total vehicle miles of travel or total vehicle emissions between 2005 and 2025 for each of the metropolitan areas listed. The projections come from conformity determinations published by the metropolitan planning organization (MPO) in each metropolitan area. For non-California areas, emissions estimates were generated using EPA's MOBILE6 emissions factor model, combined with travel estimates from a travel-demand model. For California, the emissions estimates were generated using the California Air Resources Board's EMFAC2002 model.

Most transportation-related VOCs come from gasoline vehicles, as diesels emit few, so the VOC projection is mainly influenced by projections for automobile emissions. However, the NOx trend is influenced roughly equally by projections for both gasoline and diesel vehicles.

The projections all come from these metropolitan areas' transportation plans and conformity determinations. Note that the number of vehicle-miles traveled is projected to increase by as much as 51 percent for the metropolitan areas shown in the graph. Nevertheless, total motor-vehicle VOCs are projected to decrease between 52 and 71 percent, while total motor-vehicle NOx is projected to decrease between 70 and 83 percent. The story is the same in all other metropolitan areas—air quality regulators and regional planners predict that in the future there will be much more driving and much less air pollution.

The regulators' and planners' projections are based on EPA's MOBILE6 emissions model, along with predictions of future growth in driving. As the graph shows, MOBILE6 predicts large declines in total VOC and NOx emissions, but not quite as large as the 80+ percent reductions we predicted above, particularly for VOC.

The official predictions are unreasonably pessimistic, because MOBILE6 overestimates likely future automobile emissions. How do we know? First, comparisons of MOBILE6 predictions with on-road measurements show that MOBILE6 overestimates the emissions of newer cars relative to older cars.[64] This means that MOBILE6 will underestimate future improvements in fleet-average emissions, because the model assumes that retiring older cars will reduce emissions less than it actually will, and assumes that modern cars are emitting more than they actually are.

Second, MOBILE6 assumes a slower improvement in VOC emissions during the last decade than was actually observed in on-road measurements and data from vehicle inspection programs.[65]

Third, MOBILE6 assumes that average automobile NOx emissions will decline in the future at about the same rate as they have during the last decade. This is implausible, because of the huge reduction in NOx emissions required by EPA's Tier 2 standards, which began phasing in with the 2004 model year. For example, under the Tier 1 standards, which were in effect for the 1994–2000 model years, cars and small-to-medium SUVs had to meet a standard of 0.4 gram/mile and large SUVs had to be under 0.7 gram/mile. Ultra-large passenger vehicles like the Hummer weren't even included in these standards, but were instead part of the higher-emitting "medium-duty" category. Pre–Tier 1 standards (for automobiles built before 1994) allowed NOx emissions topping 1 gram/mile.

In contrast, Tier 2 lowered allowable emissions down to a fleet average of 0.05 gram/mile. Furthermore, even the largest passenger cars, such as the Hummer, are included in the Tier 2 "light-duty" category. Tier 2 also requires vehicles to comply with the Tier 2 emission caps up to 120,000 miles, in contrast to 100,000 miles under Tier 1. Thus, rather than continuing at the same rate as in the past, the decline in automobile NOx emissions will accelerate as more Tier 2 vehicles enter the fleet and pre-Tier 2 vehicles are scrapped.

Fourth, a comparison of MOBILE6's estimates with the actual Tier 2 requirements also suggests that the model understates future improvements. According to the model, fleet-average VOC + NOx emissions are currently around 1.8 grams/mile, but will decline only to about 0.5 gram/mile by 2025, when virtually the entire fleet will be Tier 2 automobiles. But based on the actual requirements of the Tier 2 standards, an all–Tier 2 fleet of the same age distribution as the current fleet will likely emit far less than this—almost certainly less than 0.2 gram/mile.[66] Taking all of these factors into account, an all–Tier 2 fleet will likely be at least 90 percent cleaner than the average automobile on the road in 2005.

Regulators and planners would predict lower emissions in the future if MOBILE6 included more realistic projections. Rather than the 50–70 percent declines in VOCs and 70–80 percent declines in NOx predicted by regulators and planners, the real improvement will be more like 85 percent for both pollutants. But even based on current official estimates, it is clear that projected growth in vehicle travel will not prevent large declines in total vehicle emissions.

Diesel soot emissions projections are not included in regional transportation plan estimates. The MOBILE6 model makes it clear, however, that regulators believe current requirements will eliminate almost all remaining diesel soot emissions. Based on MOBILE6 projections, the soot emissions rate from diesel trucks will decrease more than 90 percent during the next twenty years or so. This is exactly what we should expect, given that we actually observe an ongoing decline in the average rate of about 8 percent per year, and that EPA's 2007 standards require a 90 percent reduction below standards that applied from 1994 to 2006.

Industrial emissions will also continue to decline. Compared with 2003, EPA's Clean Air Interstate Rule (CAIR) requires power plants to reduce total sulfur dioxide emissions 42 percent by 2010 and 64 percent by 2020.[67] The ultimate cap, which would be achieved a few years after 2020 when all banked emissions are depleted, is 73 percent below 2003 emissions. CAIR also requires additional NOx reductions. Compared with 2003, CAIR requires total annual power plant NOx to decline 42 percent by 2009 and 48 percent by 2015. Total ozone-season NOx must decline 34 percent by 2009 and 40 percent by 2015. All of these are over and above the substantial reductions in power plant emissions achieved through 2003 (see figures 4-9 and 4-10 above). EPA's Clean Air Mercury Rule will reduce power plant mercury emissions by 21 percent in 2010 and 70 percent in 2018.[68]

We saw earlier that EPA's MACT rules have already eliminated most emissions from other industrial sources besides power plants. But additional MACT rules will come into effect during the next few years, ensuring further reductions from the industrial sector.[69]

Conclusion

Compared to peak levels in the 1960s and '70s, the vast majority of air pollution emissions have already been eliminated. Already-adopted requirements will eliminate most remaining emissions during the next two decades. Air pollution has thus already been solved as a long-term problem. Remaining pollution is a near-term problem that will progressively disappear as existing requirements come to fruition over the next several years.

5

Exaggerating Air Pollution Levels; Obscuring Positive Trends

Between 2003 and 2005, the fraction of ozone monitors violating the 8-hour standard plummeted from 43 percent to 18 percent. These levels were the lowest of any three-year period on record, and all three years had lower ozone than any previous year. Though this should have been cause for celebration, the environmental group Clean Air Watch proclaimed shortly after the 2005 ozone season ended that "Smog Problems Nearly Double[d] in 2005."[1] Pennsylvania's Department of Environmental Protection warned, "Number of Ozone Action Days Up from Last Year."[2] And EPA's New England regional office noted that "New England Experienced More Smog Days during Recent Summer."[3] Writing on 2005 ozone levels in Connecticut, a *New York Times* headline lamented, "A Hot Summer Meant More Smog."[4]

Indeed, ozone levels were higher in 2005 when compared with 2004; 2005 was only the *second*-lowest ozone year since the 1970s, while 2004 was the lowest. Levels were so improbably low in 2004 that it would have been astounding if they *weren't* higher in 2005. The real news was that 2005 was one of the hottest years on record—a condition which favors high ozone—yet ozone remained at historic lows.[5] $PM_{2.5}$ and other pollutants also continue to reach new historic lows and continue to decline, yet this success has not been reported in the media, and most Americans remain unaware of it.

In the first four chapters, we showed that air pollution has steadily declined and will continue to decline for decades to come. Yet most Americans are unaware of the nation's stellar progress on air quality. Rather, according to public opinion polls, many or even most Americans believe air pollution has been getting worse and will continue to worsen in the future.

Polls also show that Americans consider environmental groups the most credible sources of information on the environment, yet these groups consistently provide inaccurate information on air pollution levels, trends, risks, and prospects, creating the appearance of a serious and worsening problem. Regulatory agencies, in whom the public also places its trust, often paint a similarly pessimistic picture.[6] In this chapter we show how this portrait is the polar opposite of reality.

Americans' Perception of Air Pollution

Americans respond pessimistically when polled about air pollution. For example, in an August 2004 poll conducted by Wirthlin for the Foundation for Clean Air Progress, 38 percent of people said they thought air pollution had gotten worse during the last decade, while 31 percent though it had stayed the same. Despite the respondents' ignorance of the nation's progress, the 2004 results were nevertheless an improvement over 2002, when a similar poll reported that 66 percent of Americans believed air pollution had gotten worse during the previous ten years, up from 61 percent in 1999.[7]

A poll commissioned by Environmental Defense in 2000 reported that 57 percent of Americans believed environmental conditions had gotten worse during the previous thirty years.[8] State-based surveys have found similar results. The Public Policy Institute of California (PPIC) reported in 2002 that 78 percent of Californians polled believed the state had made only "some" or "hardly any" progress in solving environmental problems.[9]

Americans also say they believe environmental quality will decline in the future. The 2000 Environmental Defense poll reported that 67 percent of Americans believed air pollution would get worse. Likewise, a March 2001 Gallup Poll found that 57 percent of Americans believed environmental quality was deteriorating.[10] A 1999 *Washington Post* poll reported that 51 percent of Americans believed pollution would greatly increase, up from 44 percent in 1996.[11] The 2002 PPIC poll reported that 79 percent of Californians believed the state would make little or no progress on environmental problems in the future.

There's an old saying that goes, "It's not the things we didn't know that hurt us; it's the things we knew for sure that turned out to be wrong." When it comes to air pollution, why do most Americans "know" so much that is not so? A key reason is that the major sources of the public's information on the environment—environmentalists and regulators—produce statistics that make air quality appear to be much worse than it actually is. Journalists complete the circle by relaying this misinformation with little or no critical review.

Inflating Air Pollution Levels

Each spring since 1999, the American Lung Association (ALA) has published *State of the Air*—one of the most influential reports on air pollution in the nation. Dozens of newspapers cover its release, and it is frequently cited around the country to justify stronger air quality regulations. But *State of the Air*'s portrayal of air pollution has little to do with the actual air quality of America's cities and towns. The report leads Americans to believe air quality in their communities is much worse than it really is.

For example, *State of the Air: 2003* claimed Los Angeles County exceeded the 8-hour ozone standard an average of thirty-five days per year in 1999–2001.[12] But the county's actual monitoring data for that period showed that even the worst site averaged only half that many exceedances, while the average site had only six exceedance days, or 85 percent less than ALA claimed. Figure 5-1 on the following page compares ALA's claim with actual data from Los Angeles County's fourteen monitoring sites.

ALA also gave a failing air quality grade to all of L.A. County, even though half the county's monitoring locations complied with both the federal 1-hour and 8-hour ozone standards. Each year, *State of the Air* likewise inflates air pollution levels for dozens of other populous counties around the United States, claiming much higher levels than ever actually occur.

Why are ALA's pollution estimates so much higher? ALA assigned an ozone exceedance day to an entire county on any day in which at least one location in the county exceeded the 0.085 ppm, 8-hour standard. Taking Los Angeles County as an example, if ozone exceeded the standard one day in Glendora and the next day in Santa Clarita, fifty miles away, the report counted two exceedance days for all 10 million people in the county, even though most might not have experienced even one day above the standard.

FIGURE 5-1

COMPARISON OF ALA CLAIM WITH THE ACTUAL NUMBER OF DAYS PER
YEAR EXCEEDING THE 8-HOUR OZONE STANDARD AT LOS ANGELES
COUNTY MONITORING LOCATIONS, 1999–2001

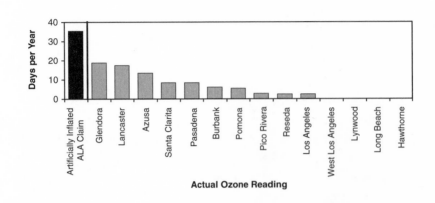

SOURCES: ALA claim is from American Lung Association, *State of the Air: 2003*, http://lungaction.org/reports/stateoftheair2003.html (accessed September 29, 2006). Actual ozone levels are from California Air Resources Board, *2006 Air Quality Data CD*, 2006, http://www.arb.ca.gov/aqd/aqdcd/ aqdcd.htm (accessed September 29, 2006).

ALA used this technique in dozens of other counties around the nation. Table 5-1 provides a few examples from *State of the Air: 2003*.

ALA continues to use this technique. According to *State of the Air: 2006*, Riverside County, California, averaged ninety days per year exceeding the 8-hour ozone standard in 2002–4.[13] But Banning, the worst location in the county, averaged fifty exceedance days, while Indio, the best location, averaged seventeen.

Regulators use a similar method of reporting ozone levels. For example, in 2001, EPA downgraded California's San Joaquin Valley (SJV) air district—a multicounty region—from "serious" to "severe" nonattainment for the 1-hour standard.[14] The change gave the region more time to attain the standard, but also required more stringent air pollution controls. In its press release on the action, EPA asserted that "air quality data from 1997 through 1999 indicate the SJV experienced 80 days of unhealthy levels of ozone air

TABLE 5-1
DAYS PER YEAR EXCEEDING THE 8-HOUR OZONE STANDARD, 1999–2001;
AMERICAN LUNG ASSOCIATION CLAIM COMPARED WITH
ACTUAL OZONE LEVELS

	Days per Year Exceeding Federal 8-Hour Ozone Standard, 1999–2001			
		Actual Ozone Monitoring Data		
County (City)	American Lung Claim Association	Worst Location in County	Average Location in County	Best Location in County
San Bernardino, CA	86	73	24	1
Fresno, CA	85	69	50	27
Harris (Houston), TX	45	22	14	5
Los Angeles	35	18	6	0
Mecklenburg (Charlotte), NC	22	18	15	10
Maricopa (Phoenix), AZ	19	8	3	0
Fairfield, CT	18	11	10	8
San Diego, CA	16	14	2	0
Dallas, TX	15	13	7	1
Philadelphia, PA	11	7	5	1
Jefferson (Birmingham), AL	11	6	5	4
Lucas (Toledo), OH	5.3	4.3	2.5	0
King (Seattle), WA	3.0	1.0	0.3	0
Clark (Las Vegas), NV	2.0	1.3	0.4	0

SOURCES: ALA's claim is from American Lung Association, *State of the Air: 2003*, http://lungaction.org/reports/stateoftheair2003.html (accessed September 29, 2006). Actual ozone levels are from analysis of site monitoring data downloaded from EPA's AIRData database, http://www.epa.gov/air/data/geosel.html (accessed September 29, 2006) and from CARB, *2006 Air Pollution Data CD*, February 2006, http://www.arb.ca.gov/aqd/aqdcd/aqdcd.htm (accessed September 29, 2006).

pollution."[15] Yet Clovis, a suburb of Fresno and the most polluted location in the SJV during the period in question, tallied forty days above the 1-hour standard—half of EPA's claim—while 40 percent of the SJV's monitoring locations actually complied with the standard.

For 2005, EPA's New England office reported that the region exceeded the 8-hour ozone standard on twenty-six days. Yet the worst location, in Danbury, Connecticut, had eleven exceedance days, and the average location had fewer than three. Much of the region never exceeded the 8-hour standard at all.

One might argue that talking about the number of days smog is elevated somewhere in a region is not misleading and paints a fair picture of the nature of the regional pollution problem. But the health effects of smog depend in part on how often a given person is exposed. Because no one is exposed to smog anywhere near as often as activists and regulators claim, they are encouraging people to overestimate vastly their risk from air pollution.

Another way to see this is to ask yourself, "Where can I find the people in New England who experienced twenty-six ozone exceedance days in 2005?" The answer is, "You can't." People in Danbury, Connecticut, were exposed to the worst ozone pollution—a total of eleven 8-hour exceedance days in 2005. Likewise, you can't find anyone in Los Angeles County who faced thirty-five 8-hour exceedance days per year during 1999–2001. The worst place you can find is Glendora, with eighteen days per year.

The Public Interest Research Group (PIRG) applied these techniques at the state level, creating an even larger difference between claimed and actual pollution levels. In its 2002 report, *Danger in the Air*, PIRG claimed that California exceeded the 8-hour ozone standard on one hundred thirty days in 2001.[16] Yet nearly half of the state's monitoring locations had no exceedance days, while the average location had seven. Even the worst location in California had only about half as many exceedances as PIRG claimed for the whole state. PIRG claimed inaccurately large ozone problems for every state it scrutinized.

PIRG also created another statistic it called "exceedances" and claimed California had 1,359 of them in 2001. PIRG simply added up all the 8-hour exceedance days at each monitor in a state and applied that sum to the whole state. By this method, even Wisconsin had 169 "exceedances," even though about two-thirds of its monitors complied with the 8-hour ozone standard and 90 percent complied with the 1-hour standard.

One hundred sixty-nine is a scary number of times to have unhealthy air, but it has nothing to do with actual pollution levels anywhere in Wisconsin. Having created a false impression that most of the United States exceeds the 8-hour ozone standard many more times per year than it actually does, PIRG went on to claim without qualification that "our cities, suburbs and even our national parks are shrouded in smog for much of the summer."[17]

In November 2005, after the United States had achieved its second-lowest ozone levels in history, Clean Air Watch claimed that "the federal health standard for ozone, or smog, has been breached 3,423 times in 40 states and the District of Columbia this year."[18] This was a year in which the average ozone monitoring location exceeded the 8-hour standard on less than three days, 43 percent of monitoring locations had zero 8-hour exceedances, and 87 percent had zero 1-hour exceedances.

Even for 2004, when the number of 8-hour exceedance days for the average monitor was 1.7, and 68 percent had zero exceedances, Clean Air Watch claimed there were 1,930 exceedances around the United States. Like PIRG, Clean Air Watch simply summed the 8-hour exceedances for every monitoring site in the country to come up with an inaccurate picture of exposure to pollution. This approach is comparable to doctors adding up the cholesterol levels of one hundred patients with normal cholesterol and concluding that Americans have cholesterol levels one hundred times greater than normal.

Journalists unwittingly abet this misrepresentation because they generally pass along regulators' and activists' claims without critical review. For example, a *New York Times* feature story on air pollution in Los Angeles claimed the L.A. area exceeded the 1-hour ozone standard on sixty-eight days in 2003.[19] In fact, the worst site in the L.A. metropolitan area had thirty-nine exceedance days that year, and the average site had about ten. Though dozens of newspapers cover the ALA, PIRG, and other reports, few include any critical analysis of the reports' assertions, and virtually none have checked the pollution numbers they are given to see if they jibe with actual monitored levels. As is apparently the case with the *Times* story, reporters seem to be unaware that they are being fed inaccurate numbers.

With the release of *State of the Air: 2004*, ALA set up its Web site to provide its inflated numbers to visitors and pass them off as actual pollution levels. ALA's home page allows visitors to type in their zip codes and receive information ostensibly about air pollution levels in their neighborhood.

But for anyone living in a county with more than one ozone monitor, the information ALA provides is false. For example, a Pasadena, California, resident who types in zip code 91101 would be told that her area had 173 "high ozone" days (days exceeding the 8-hour ozone standard) during 2001–3, or nearly 58 per year.[20] But inspection of actual monitoring data for Pasadena shows that there were only sixteen 8-hour exceedance days per year during 2001–3. ALA's number is too high by a factor of 3.6.

Someone living in the Lynwood area of Los Angeles (zip code 90280) would be given the same fifty-eight-days-per-year figure, as would someone living in Santa Clarita (zip code 91350). Yet these two areas experienced, respectively, zero and forty-nine days per year exceeding the 8-hour ozone standard. All three of these cities are in Los Angeles County, and Santa Clarita had the worst ozone in the county during 2001–3. But it doesn't matter where in Los Angeles County you live. No matter what L.A. County zip code you type in, ALA's Web site will always give you the same pollution number.

Here's another example, this time for Harris County, Texas, where Houston is located: If you type in the zip code 77074, in southwest Houston, ALA will tell you there were ninety-nine 8-hour ozone exceedances during 2001–3, or thirty-three per year. But the monitoring site in that zip code recorded only fifteen exceedance days per year, less than half ALA's claim; and this is the worst location in Harris County.[21] If you type in 77003, the zip code for central Houston, ALA gives you the same answer—thirty-three 8-hour exceedance days per year. But the monitoring site in that zip code averaged only five exceedances per year, making ALA's claim too high by more than a factor of 6.[22] Harris and Los Angeles counties are not isolated examples, but represent the norm.

ALA's Web site asks, "Are you one of the 152 million Americans breathing unhealthy air?"[23] Not only is this a great exaggeration of the total number of people living in areas that violate one or more EPA air pollution standards (see chapter 6 for more on this), it gives the impression that all of these people often or regularly breathe "unhealthy air," even though the air exceeds EPA's 8-hour ozone standard less than 1 percent of the time in most areas that violate the standard. Furthermore, as we will see in chapter 7, the 8-hour standard is set at such a low level that exceeding the standard has, at worst, minor health implications.

ALA's claims are even more inflated for $PM_{2.5}$ than for ozone. The 2004 edition of *State of the Air* included $PM_{2.5}$ data for the first time, and ALA not only used the system described above for counting the number of $PM_{2.5}$ exceedances in a county; it also used a standard that is much tougher than the federal standard. EPA set its 24-hour standard at 65 $\mu g/m^3$.[24] However, ALA counts an exceedance on any day in which $PM_{2.5}$ is greater than a far lower level of 40 $\mu g/m^3$. As a result, even though only ten U.S. counties had a monitoring location that violated EPA's 24-hour standard as of 2002 (and only four counties as of 2003), ALA gave failing $PM_{2.5}$ grades to 107 counties in *State of the Air: 2004*. For example, Cook County (Chicago), Illinois, which had zero days exceeding EPA's 24-hour standard from 2000 to 2002, was cited by ALA for unhealthy $PM_{2.5}$ levels on forty-three days.[25] Hundreds more counties received grades of B, C, or D, even though they complied with the standard with plenty of room to spare.

How can ALA justify this choice? They had help from EPA. In order to simplify air pollution standards for public consumption, EPA has created an "Air Quality Index" (AQI). The AQI puts different kinds of pollution on the same scale with the goal of making interpretation of pollution levels more straightforward. Thus, the AQI is standardized so that a given range of values always means the same thing. For example, an AQI between 50 and 100 is dubbed "moderate," while an AQI between 100 and 150 is dubbed "unhealthy for sensitive groups." EPA considers an AQI above 150 to be "unhealthy" for everyone.

Different pollutants have different standards and are found in air at different levels, so EPA has created a conversion system for each pollutant between actual pollution levels and these standardized AQI values. Normally, the concentration level equal to a given pollution standard is set to an AQI value of 100. For example, for ozone, a value of 0.085 ppm is pegged to an AQI value of 100, while 0.105 ppm equals an AQI of 150. Or to put it another way, EPA considers small exceedances of a standard to be "unhealthy for sensitive groups." Larger exceedances receive a blanket "unhealthy" rating.

However, EPA departed from this system when it assigned AQI values for 24-hour $PM_{2.5}$ levels. Rather than setting 65 $\mu g/m^3$ $PM_{2.5}$, the level of the 24-hour $PM_{2.5}$ standard, to an AQI of 100, EPA set 40 $\mu g/m^3$ to an AQI of 100, and 65 $\mu g/m^3$ to an AQI of 150.[26] In other words, even levels well below EPA's standard will trigger an "unhealthy for sensitive groups" warning.

The case of the AQI for $PM_{2.5}$ illustrates the symbiotic relationship between regulatory agencies and activist groups. EPA had to go through a transparent legal and administrative process in order to develop and implement its $PM_{2.5}$ standards. But after setting standards intended to protect Americans "with an adequate margin of safety," EPA then created a sort of shadow standard for its public descriptions of $PM_{2.5}$ pollution that is far more stringent than the legal standard it established to determine compliance with the Clean Air Act and federal health standards.

Since the shadow standard is used to trigger daily air pollution health advisories around the country, EPA created the means for state regulators and activists to create the appearance of continuing widespread air pollution danger, even though more than 99 percent of $PM_{2.5}$ monitoring sites already attain the 24-hour standard. For example, 64 percent had at least one reading between 2001 and 2003 that was greater than 40 $\mu g/m^3$, but only 8 percent had at least one reading that was greater than 65 $\mu g/m^3$.[27] Thus, without changing the 24-hour standard, EPA was able to increase the geographic extent of $PM_{2.5}$ "code orange" warnings by a factor of 8, as well as greatly increase the number of days per year on which such warnings would be issued. EPA also created a superficially plausible justification for organizations like ALA to give out failing grades to dozens of counties that nevertheless comply, usually by large margins, with EPA's 24-hour $PM_{2.5}$ standard. Using 40 $\mu g/m^3$ as an alternative $PM_{2.5}$ standard caused ALA to exaggerate greatly the extent of 24-hour $PM_{2.5}$ exceedances in the 2004, 2005, and 2006 editions of its report.

In September 2006, EPA adopted a new 24-hour standard set at 35 $\mu g/m^3$.[28] ALA used this new standard for the 2007 edition of its report, so its $PM_{2.5}$ exceedance count now suffers only from the same inflation problem as the ozone exceedance count.

Counting Clean Areas as Polluted

If one monitor in a county exceeds a pollution standard, both EPA and ALA count everyone in the county as breathing air that exceeds the standard.

FIGURE 5-2

EIGHT-HOUR OZONE LEVELS RELATIVE TO THE FEDERAL STANDARD AT ALL
MONITORING SITES IN FOUR LARGE COUNTIES, 1999–2001

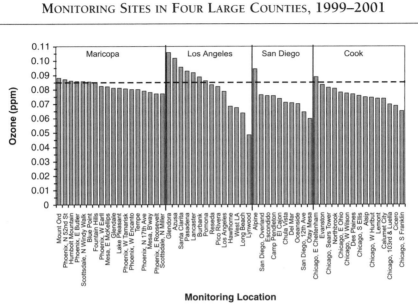

SOURCE: Ozone levels are from EPA's AIRData database, http://www.epa.gov/air/data/geosel.html (accessed
September 29, 2006).

NOTES: For each monitoring location, the ozone level is the average of the fourth-highest 8-hour ozone for
1999, 2000, and 2001—the same value EPA uses to determine attainment of the 8-hour ozone standard.
A site violates the standard if this value is equal to or greater than 0.085 ppm, as marked by the horizontal dotted line. Phoenix is located in Maricopa County, and Chicago is located in Cook County.

However, as shown in figure 5-2, in some cases almost all of a county has
clean air, and only one or a few locations violate the standard. The chart displays 8-hour ozone levels (bars) relative to the 8-hour standard (dotted
line) in four of the six most populous counties in the country.[29]

Note that in Cook and San Diego counties, only one monitoring location
violated the 8-hour standard in 2001, while even in Los Angeles County only
half of all monitoring locations violated the standard. Nearly 21 million
Americans live in these four counties alone; ALA and EPA claimed incorrectly
that 80 percent of them live in areas that violate federal ozone standards.[30]
Nor are these isolated examples. Many counties that received a failing grade

for air quality from ALA nevertheless complied with federal pollution standards over much or most of their land area. Counting clean areas as dirty causes EPA and environmental activists to erroneously include tens of millions of Americans as living in areas that violate federal pollution standards—a fact that will be demonstrated in greater detail in the next chapter.

The "official" numbers put out by EPA create a patina of legitimacy for environmental activists' frightening claims as well. For example, when EPA announced on April 15, 2004, its estimate of 159 million Americans breathing unhealthy air, a Sierra Club spokesman said the numbers "show that more than half of all Americans live in places with unhealthy air . . . [that] Americans face a serious air pollution problem."[31] This is an exaggeration both of the number of people in areas that violate the 8-hour standard and of the risk from such violations.

Most areas that violate the 8-hour standard spend no more than about 50 hours per year with ozone greater than 0.085 ppm. And hardly any areas that exceed the 8-hour standard ever reach the higher ozone level of the 1-hour standard. Those areas that do violate the 1-hour standard, typically spend only two to three hours per year with ozone greater than 0.125 ppm. Calling fifty hours per year with ozone a few parts per billion above the moderate level of the 8-hour standard, or even a few hours per year with ozone above the higher level of the 1-hour standard a "serious" pollution problem, is a stretch. Furthermore, as will be shown in chapter 7, even EPA's and the California Air Resources Board's own estimates conclude that reducing ozone from recent levels down to the 8-hour standard, or even well below the standard, would at best confer tiny and virtually unnoticeable health benefits.

Same Failing Grades for High- and Low-Pollution Areas

Activists also blur the distinction between areas that barely exceed EPA standards and those that exceed them by large margins. Since the vast majority of violating areas fall into the former category, this allows activists to create the impression that much of the country frequently has dangerous air. For example, ALA gives a failing air quality grade to any county that averages at least 3.3 days per year exceeding the 8-hour ozone standard.

This means that many areas that spend 100 percent of the year with ozone below the 1-hour standard and 99.5 percent below the 8-hour standard get the same F grade as the worst areas of California, which have ten or twenty 1-hour exceedances days and dozens of 8-hour exceedances days per year.

For example, in *State of the Air: 2004*, Phoenix, the Bronx, Salt Lake City, Albany, Evanston, Tampa, and Waukesha all received the same F grade as the worst areas of California. Yet these other areas complied with the 1-hour ozone standard and either complied with or barely exceeded the 8-hour standard over most or all of their land area. Indeed, it is not uncommon for ALA to give F grades to cities that comply with both the 1-hour and 8-hour standards over much or all of their land area.

ALA's pollution rankings also create a false impression of high pollution in many areas of the country. Each year, ALA ranks the worst twenty-five metropolitan areas for air pollution. For example, in *State of the Air: 2004*, Knoxville, Tennessee, was ranked ninth in the nation. That sounds pretty bad. But, as shown in figure 5-3 on the following page, once you get past the worst few areas in the country—all in the Los Angeles and San Joaquin Valley areas of California—pollution levels are relatively low everywhere else. Figure 5-3 displays the actual number of 8-hour and 1-hour ozone exceedances days per year at the worst location in each of ALA's twenty-five worst cities. We used data for 2000–2, the same years used by ALA for the 2004 edition of its report.

Note that even though ALA ranked Knoxville ninth-worst in the country for 2000–2, it had virtually no 1-hour ozone exceedances and twenty-one 8-hour exceedance days per year. And even those numbers represent only the worst of Knoxville's six monitoring locations.[32] The average and best Knoxville locations averaged, respectively, 12.3 and 1.3 8-hour exceedance days per year.

And so it goes for other metro areas on ALA's list. Outside of a few California areas, even the worst areas in the country have relatively low ozone levels. But the fact that an area can violate an EPA standard by exceeding a given pollution level just a few days per year makes it easy for advocacy groups to translate this standard into simplistic letter grades that make Pittsburgh sound as bad as San Bernardino, when it doesn't even come close. By ignoring context, ALA uses its rankings to make air pollution appear much worse than it actually is throughout the country.

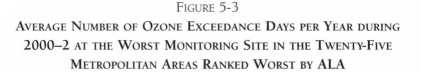

FIGURE 5-3

AVERAGE NUMBER OF OZONE EXCEEDANCE DAYS PER YEAR DURING
2000–2 AT THE WORST MONITORING SITE IN THE TWENTY-FIVE
METROPOLITAN AREAS RANKED WORST BY ALA

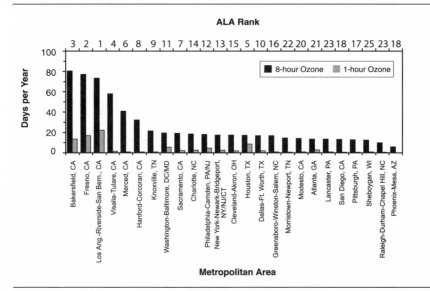

SOURCES: American Lung Association, *State of the Air, 2004*, May 1, 2004, http://lungaction.org/reports/sota04
_full.html, and analysis of site monitoring data downloaded from EPA's AIRData database, http://www.
epa.gov/air/data/geosel.html (accessed September 29, 2006).

NOTES: Data are averages for 2000–2, the same years used by ALA for *State of the Air: 2004*. Metropolitan
areas are ranked from highest to lowest ozone levels based on the location in each metro area with the most
8-hour exceedance days per year. Numbers along the top of the graph give ALA's ranking for each area.

Note also that ALA's rankings, which appear along the top of figure 5-3,
are different from our rankings. Our rankings are based on the actual number of 8-hour exceedances at the worst ozone monitoring site in each metro
area. But ALA's are based on its artificially inflated ozone values. ALA's
method inflates the values by different percentages in different areas,
depending on how many monitors a region has, and on how much ozone
levels vary from place to place within a given region. Thus, ALA not only
exaggerates air pollution levels; it also ends up with rankings that have little
relationship to actual relative pollution levels among cities.

Air Quality: Much Worse on Paper than in Reality

A few areas of California have by far the highest air pollution levels in the United States, but according to dozens of news stories and editorials over the last several years, most other cities in the United States are the ones with "some of the worst air pollution in the nation."

Here are a few among many examples. According to the *Chicago Sun-Times*, Chicago has "some of the worst air pollution in the nation."[33] The Dallas-Fort Worth area has "some of the country's worst air," claims the *Fort Worth Star-Telegram*.[34] The *Baltimore Sun* says Baltimore has "some of the worst air pollution in the country," as well.[35] The New York metropolitan area? "Some of the country's dirtiest air," according to the *Westchester Journal News*.[36] Atlanta, according to the *Atlanta Journal-Constitution*, has "some of the worst air pollution in the country."[37] The *Washington Post* puts not only the Washington-Baltimore metropolitan area but also Phoenix in the "some of the worst air pollution" fraternity.[38]

Sometimes it is entire states that have "some of the worst air pollution." New Jersey, the *Bergen Record* says, has "some of the worst air pollution in the country."[39] But just across the Hudson River the *New York Times* claims not only that the state of New York "has some of the nation's dirtiest air," but that "the smog in Connecticut is among the worst in the nation."[40] Tennessee experiences "some of the worst air pollution in America," according to the *Chattanooga Times Free Press*.[41] Maryland is "faced with some of the worst air pollution in the country," according to the *Baltimore Sun*.[42]

The citations above come from journalists and editors, who get much of their information from environmental activists. But activists also make many "some of the worst" claims directly. For example, the New Jersey Public Interest Research Group claims that "Passaic County [New Jersey] suffers from some of the worst air pollution in the country."[43] North Carolina PIRG says North Carolina "has some of the worst air quality in the country."[44] Ohio PIRG claims it is Ohio that has "some of the worst air pollution of any state."[45] Tennessee Conservation Voters claims Tennessee has "some of the worst air pollution" in the country."[46] The New York chapter of the American Lung Association impugns New York State's air as being "on par with some of the worst polluted air in the country."[47]

This is just a small sample of the dozens of news stories and press releases in which environmentalists have claimed some area or other has some of the worst air pollution in the country.[48] Even without looking at any data, it's clear they can't all be right. And, in fact, all of them are wrong, by a large margin.

The director of the American Lung Association's Santa Clara, California, chapter didn't even bother with a vague "some of the worst" claim. Instead he asserted, "We've got the same smog problems in the [San Francisco] Bay Area that they have in Los Angeles."[49] Figure 5-4 shows that it would be hard to make a more erroneous statement about the relative air quality of Southern California and the Bay Area.

Refuting Their Own Claims

Activists' reports sometimes contain information that refutes the reports' own claims. For example, in *Clearing the Air with Transit Spending*, the Sierra Club claimed areas that spend more per capita on transit have less air pollution, and graded metropolitan areas from A through F based on how much they spent on transit and on the amount of motor vehicle pollution emitted per capita in each area. But the Sierra Club report does not provide evidence that more transit spending or a better grade has any relationship to actual air quality. They don't. The correlation between cities' per-capita transit spending and the number of days per year they exceed the 8-hour ozone standard is literally *zero*.

This is exactly what one should expect, since outside of the New York metropolitan area, transit's market share ranges from 0 to 5 percent of all person-miles of motorized travel, no matter how much cities spend on it.[50] With so few people using transit anywhere, it simply couldn't have a measurable effect on air quality. An even greater irony is that the cities that got D and F grades from the Sierra Club actually had *better* air quality than areas that received a grade of B and C (no city got an A).[51]

In *More Highways, More Pollution*, PIRG claims that "building new highways will do little to alleviate traffic congestion in the long run and likely will exacerbate already severe air pollution problems in metropolitan areas across the country."[52] We showed earlier that this is just the opposite of

FIGURE 5-4

COMPARISON OF 8-HOUR OZONE EXCEEDANCE DAYS DURING 2000–2
IN THE SAN FRANCISCO BAY AND LOS ANGELES AREAS

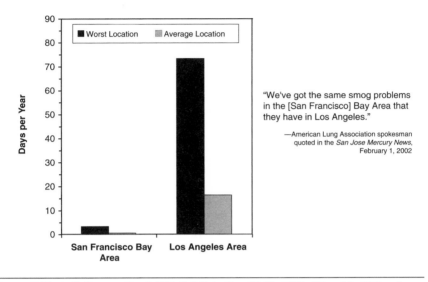

"We've got the same smog problems in the [San Francisco] Bay Area that they have in Los Angeles."

—American Lung Association spokesman quoted in the *San Jose Mercury News*, February 1, 2002

SOURCE: California Air Resources Board, *2003 Air Pollution Data CD*. The 2003 edition of the CD is no longer available. The 2007 edition, which includes all data on the 2003 edition, plus data collected since then, is available at http://www.arb.ca.gov/aqd/aqdcd/aqdcd.htm (accessed March 21, 2007).

NOTE: The chart compares the worst and average areas in the Bay Area Air Quality Management District and the South Coast Air Quality Management District (the Los Angeles metropolitan region).

what has happened during the last few decades (see figure I-1 in the Introduction). Between 1980 and 2005, total miles of driving doubled, while air pollution dropped between 20 and 96 percent (depending on the pollutant and the statistic by which it is measured).[53]

But PIRG's report suffers from an even more serious flaw. PIRG does not provide any evidence on whether the fundamental premise of its report—that areas with more road-miles per capita have higher air pollution levels—has any real-world validity. PIRG includes an estimate of vehicle NOx and VOC emissions per capita for metropolitan areas around the United States and claims that those with more road-miles per capita have more emissions and therefore more ambient pollution. If this chain of logic were correct, it might bolster PIRG's contention. However, the air pollution monitoring data don't bear this out.

As shown in figure 5-5, there is no correlation between PIRG's estimate of vehicle emissions per capita and actual measured ozone levels. Several factors could explain this. First, PIRG used EPA's emissions inventory for estimating emissions per capita, even though field studies suggest motor vehicles make a substantially larger contribution to total VOC emissions (see page 54 for details on emissions sources) than regulators' official inventories claim.

Second, pollution levels vary based on differences in weather from place to place, and weather variation probably has a larger effect on air quality than differences in emissions. Third, urban form affects pollution levels. Higher population density means more emissions per unit area. Suburbanization is associated with a slight increase in total driving per capita, but it reduces the density of emissions much more, because population density has only a minor effect on how much people drive.[54] Thus, overall a suburban development pattern reduces emissions per unit area, and therefore reduces ambient pollution levels at any given location.

Ambient pollution levels are what matter for health. Figure 5-5 shows that there's no relationship between metropolitan areas' per-capita emissions and ozone levels. PIRG's entire study is thus based on an invalid premise.

Making Pollution Decreases Look like Pollution Increases

In 2002, a report commissioned by the Rockefeller Family Fund's Environmental Integrity Project highlighted the fact that some power plants increased their sulfur dioxide emissions between 1990 and 2001 and noted that "sulfur and nitrogen emissions from power plants significantly increase the amount of fine particles in the air."[55]

The Rockefeller report implies that sulfate particulate levels increased, but omits any actual data on trends in ambient sulfate levels. In fact, sulfate levels declined not only on average, but at virtually every site in EPA's monitoring network.[56] Figure 5-6 on page 102 compares average sulfate levels during 1989–91 (triangles) or 1997–99 (circles) with levels

FIGURE 5-5

RELATIONSHIP BETWEEN PER CAPITA MOTOR VEHICLE EMISSIONS AND NUMBER OF OZONE EXCEEDANCE DAYS PER YEAR IN U.S. METROPOLITAN AREAS

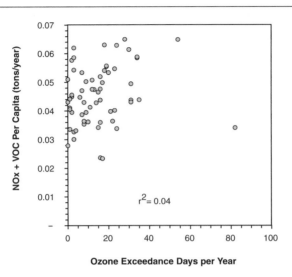

SOURCE: Motor-vehicle emissions per capita is from PIRG, *More Highways, More Pollution*. Ozone exceedance days are from monitoring data downloaded from EPA's AIRData database, http://www.epa.gov/air/data/geosel.html (accessed September 29, 2006).

NOTES: Each point represents a single metropolitan statistical area (MSA). "NOx + VOC per capita" is for 1999. All MSAs with population more than one million are included. "Ozone exceedance days" is the number of 8-hour ozone exceedances in 1999 at the worst location in each MSA.

during 2003–5 at all monitoring sites with continuous data during the given period. Points that fall below the diagonal line signify locations where sulfate levels decreased.

Note that sulfate decreased virtually everywhere during both periods, and that the sites that started out with the highest levels experienced the greatest improvements, contrary to what the Rockefeller Fund report implies.[57] As shown earlier in chapter 4, overall power plant emissions have been steadily decreasing under the Clean Air Act's Title IV sulfur dioxide and NOx programs, and this has been reflected in lower particulate sulfate levels measured around the nation. If power plant pollution were increasing, this is just the opposite of what we would have expected.

FIGURE 5-6

COMPARISON OF AVERAGE SULFATE PARTICULATE MATTER LEVELS
DURING **1989–91** OR **1997–99** WITH **2003–5** FOR ALL SITES WITH
COMPLETE DATA FOR THE GIVEN TIME PERIODS

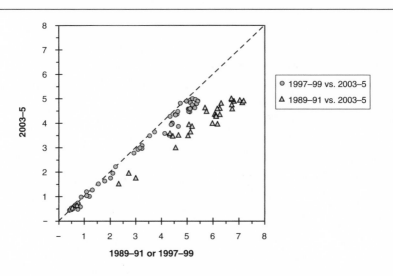

SOURCE: Sulfate monitoring data downloaded from EPA's CASTNET data Web site, www.epa.gov/castnet (accessed March 21, 2007).

NOTES: Each point represents an individual monitoring location. Points below the diagonal line represent areas where sulfate decreased between 1989–91 and 2003–5 or between 1997–99 and 2003–5.

Other publications on power plant emissions run in the same vein as the Rockefeller Fund's report. These include PIRG's *Darkening Skies: Trends Toward Increasing Power Plant Emissions*, put out in 2002, and *Power to Kill* and *Death, Disease, and Dirty Power*, by the Clean Air Task Force. In 2005, PIRG released *Pollution on the Rise: Local Trends in Power Plant Pollution.*[58]

Readers of these reports would never know that PM levels have been dropping and will continue to drop. For example, PIRG's *Darkening Skies* reports that three hundred power plants increased their SO_2 emissions between 1995 and 2000. But PIRG never mentions that total power plant SO_2 emissions declined 29 percent from 1990 to 2000, and that federal law at the time *Darkening Skies* was published required an additional 20 percent SO_2 reduction from power plants between 2000 and 2010.[59] More importantly, actual SO_2 levels in air have steadily declined across the United States for as long as SO_2 has been measured (see figures 1-4 and 1-5 in chapter 1), and virtually the entire nation attains federal SO_2 standards, almost always with plenty of room to spare.

PIRG also fails to mention that sulfate PM levels across the eastern United States declined, often by a large percentage, at every single monitoring location between 1990 and 2001, due to ongoing SO_2 reductions (see figure 3-2 in chapter 3 for trends in particulate sulfate levels).[60] Indeed, *Darkening Skies* contains no information at all on actual trends in pollutant emissions or actual PM levels in any community, despite the wealth of data available from hundreds of monitoring locations in populated areas around the country.

Another technique for creating the false impression that air pollution is rising is to put out alarming press releases in years when pollution rises compared to the previous year, but to keep mum in years when pollution falls. This is what the Clean Air Trust did late in 2002 when it proclaimed, "New Survey Finds Massive Smog Problem in 2002."[61] PIRG also highlighted the rise in ozone from 2001 to 2002 in *Danger in the Air*. Ozone levels did, indeed, rise between 2001 and 2002 (see chapter 2 above). However, the Clean Air Trust did nothing to draw attention to sharp drops in air pollution from 1999 to 2000, 2002 to 2003, and 2003 to 2004.

Selective attribution of the causes of air pollution trends also creates a misleading appearance of deteriorating air quality. When ozone is higher than average, activists blame it on emissions; when it is lower than average, they blame it on favorable weather.[62] In fact, in all years—good, bad, or average—weather is the main factor driving year-to-year variations in the number of ozone exceedance days. All else equal, years with more sun, less rain, and less wind will have higher ozone levels. The zigzagging in the trend line in figure 2-1 of chapter 2 is due to these annual variations in meteorology. But as this figure also shows, the long-term trend is downward, and this improvement is due to ongoing declines in emissions of ozone-forming pollutants.

As shown in chapter 2, in most of the United States, 2004 had by far the lowest ozone levels ever measured. The average U.S. monitoring location exceeded the 8-hour standard on just 1.7 days in 2004—less than half the average for 2003, which was itself a record-low year. Nevertheless, activist groups continued to release gloomy air quality reports during these two years.

When the improvement was acknowledged, it was attributed to the weather.[63] And, indeed, the stunningly low ozone levels in 2004 would not have been possible without help from an atypically cool and wet summer. Nevertheless, other years have had weather unfavorable to ozone formation, but none had ozone levels anywhere near as low as 2004.

Environmentalists and journalists have also failed to note that 2003, 2004, 2005, and 2006 were the four lowest ozone years on record. If high temperatures were all that mattered, the meteorology of 2005 and 2006 should have resulted in high-ozone years.[64] Not only have activists failed to explain why these hot years had low ozone, they've failed to even mention that they were low-ozone years in the first place. The low ozone levels of the last few years would not have been possible without ongoing reductions in ozone-forming emissions.

Air pollution of all kinds has steadily declined despite large increases in driving, energy production, and economic activity. *State of the Air: 2004* claimed that "the Clean Air Act is seriously at risk."[65] The 2005 installment included a section headed, "The 'Endangered' Clean Air Act."[66] Meanwhile, air pollution levels have continued downward to their lowest levels in history. As detailed in chapter 4, even after accounting for growth, requirements that have already been adopted or implemented will eliminate almost all remaining human-caused air pollution during the next twenty years or so.

Climate Change and Air Pollution

Because, all else being equal, higher temperatures are associated with higher ozone levels, environmentalists and many scientists and government officials are claiming that air pollution will increase in the future due to climate change. Two years ago, for example, the Natural Resources Defense Council (NRDC) predicted that by 2050, increasing temperatures would cause a 50 percent rise in days exceeding the federal 8-hour ozone standard each year.[67] The federal government's 2002 *Climate Action Report* also cited potential increases in air pollution due to higher temperatures.[68] The attorney general of Massachusetts put ozone increases near the top of his list of harms from global warming.[69]

Even academic and government scientists have entered the debate as activists. NRDC's *Heat Advisory* was written by public health professors from Johns Hopkins and Columbia University, along with atmospheric and environmental scientists from Yale, the State University of New York at Albany, the University of Wisconsin, NASA, and the U.S. Department of Agriculture. Some of these scientists also published their *Heat Advisory* results in the scientific journal *Environmental Health Perspectives* (*EHP*).[70]

What these studies fail to take into account is that a roughly 1°F temperature rise during the last thirty years was accompanied by sharp nationwide declines in ozone levels and levels of all other air pollutants.[71] Where about 80 percent of monitors violated the federal 8-hour ozone standard during the late 1970s, only about 15 percent do so today. The average number of 8-hour ozone exceedance days per year has declined more than 85 percent. At worst, warming will cause ozone to decrease slightly less than it otherwise might have.

As a result of ongoing declines in ozone-forming emissions, U.S. ozone levels are becoming less sensitive to high temperatures. That is, high temperatures are less likely to be associated with an ozone exceedance today when compared with ten or twenty years ago. Figure 5-7 on the following page shows the trend in the ratio of days per year exceeding federal ozone standards to days per year with temperature greater than 90°F. The graph is based on ozone and temperature data for ten metropolitan areas chosen for their relatively high ozone levels and regional representation. Between 1982 and 2005, the likelihood of an 8-hour ozone exceedance on a hot day dropped 73 percent, while the likelihood of a 1-hour exceedance dropped 93 percent.[72]

So on what did NRDC and its allied university and government scientists base their predictions of ozone increases? They assumed that ozone-forming emissions in 2050 would be the same as they were in 1996. Yet by the time *Heat Advisory* and the *EHP* journal article were published in 2004, VOC and NOx emissions had already declined well below their 1996 levels (see chapter 4). So the study was based on emissions levels that were already eight years out of date by the time it was published. In actuality,

FIGURE 5-7

TREND IN THE RATIO OF DAYS PER YEAR EXCEEDING FEDERAL OZONE
STANDARDS TO DAYS PER YEAR WITH TEMPERATURE GREATER THAN 90°F

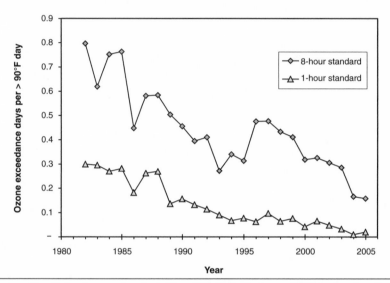

SOURCES: Air pollution data were downloaded from EPA's Air Quality System (AQS) database, http://www.epa.
gov/ttn/airs/airsaqs/detaildata/downloadaqsdata.htm and http://www.epa.gov/ttn/airs/airsaqs/archived%
20data/downloadaqsdata-o.htm (accessed November 27, 2006). Temperature data were downloaded from
the National Climatic Data Center, Summary of the Day (Data Set TD-3200), http://ncdc.noaa.gov (accessed
October 3, 2006).

NOTES: Based on ozone and temperature data for ten metropolitan areas: Atlanta, Baltimore, Charlotte,
Chicago, Cincinnati, Houston, Los Angeles, Nashville, New York, and Philadelphia. Ozone exceedances
were calculated as the average for all monitoring sites in an area with continuous data. The year 1982 was
the earliest time period for which all of the cities had at least one continuously operated monitoring site.

long before 2050, ozone-forming emissions will have declined at least
80 percent below current levels. At best, these scientist-activists estimated
how much higher ozone levels would have been back in 1996 if temperatures had been a few degrees warmer.

Ironically, temperature isn't necessarily the climate variable with the greatest effect on ozone levels. Wind speed and mixing height—the height to
which air and pollutants can mix upward from the ground—are probably
more important. Gentler winds and lower mixing heights allow ozone to

build up. Higher temperatures are associated with lower winds, but also with higher mixing heights. It's not clear which effect would dominate.

Rainfall also affects ozone. More frequent rains would tend to reduce the number of high-ozone days. Precipitation has increased in the United States during the last century, and climate models predict it will increase further if the planet warms.[73] *Heat Advisory* and the *EHP* article don't assess the potential effects of these other meteorological changes on ozone levels, so even with the inflated levels of ozone-forming emissions, they may be overstating the resulting ozone levels. Massachusetts's attorney general also fails to account for the effects of meteorology on future ozone. His Web site claims that warming will increase summer rainfall 10 percent by 2100 in Massachusetts, but he doesn't make the connection that more rain could offset the ozone-increasing effect of higher temperatures.

While climate activists and scientists have focused on the potential effects of global warming on ozone levels, another likely result of warmer temperatures would be a *reduction* in $PM_{2.5}$ levels. About one-half to two-thirds of all $PM_{2.5}$ is made up of "semi-volatile" nitrates and organic compounds.[74] These compounds evaporate as temperature rises, reducing $PM_{2.5}$ levels. A recent study of the Los Angeles metropolitan region concluded that, all else equal, a 9°F increase in temperature would reduce peak $PM_{2.5}$ by 25 percent.[75]

NRDC's *Heat Advisory* report cites this Los Angeles study, but both misstates and sidesteps its $PM_{2.5}$ results. *Heat Advisory* claims,

> Ozone and nonvolatile secondary PM will generally increase at higher temperatures because of increased gas-phase reaction rates. Interannual temperature variability in California, for example, can increase peak O_3 [ozone] and 24-hour average $PM_{2.5}$ by 16 percent and 25 percent, respectively, when other meteorological variables and emissions patterns are held constant.[76]

In fact, the Los Angeles study concluded that, all else equal, higher temperatures increase ozone but *decrease* $PM_{2.5}$. *Heat Advisory's* claim of a 25 percent increase in $PM_{2.5}$ is a misreporting of the 25 percent $PM_{2.5}$ *decrease* that the Los Angeles study actually found.

The epidemiological studies cited by EPA and environmental activists to support more stringent air pollution regulations have linked $PM_{2.5}$ to far more widespread and severe health effects than ozone, including much higher rates of premature mortality (we critique these health claims in chapter 7). EPA attributes more than 90 percent of the health benefits of the Clean Air Act to PM reductions and less than 1 percent to ozone reductions.[77] Of course, just as for ozone, already-adopted requirements will eliminate most remaining human-caused $PM_{2.5}$ during the next two decades. Whatever the effects of warming on air pollution, they will be irrelevant if and when significant warming occurs. But this doesn't change the fact that within the activists' own paradigm—a paradigm that assumes great harm from current levels of air pollution, no change in future pollution emissions, and consideration only of the effect of temperature among all meteorological variables—the net effect of global warming would be a *reduction* in harm from air pollution—just the opposite of what NRDC and its allies in government and academia have claimed.

Conclusion

These practices have disturbing implications. Activists, regulators, and journalists who claim that virtually everyone breathes "some of the worst air pollution in the country" and that it is only getting worse create an unwarranted climate of fear and anxiety.

Journalists should be acting as a check on these misleading portrayals, but they are instead a part of the problem. Perhaps journalists, like much of the public, consider environmentalists and regulators to be the presumptive guardians of the public good and see scientists as relatively objective analysts whose research methods and reports are unaffected by ideology or politics. Reporters get most of their information on air pollution levels and trends from activists', regulators', and scientists' reports and press releases. News stories suggest that journalists generally take this information at face value.

These inaccuracies have now been repeated so often that they have become "common knowledge." Many news stories do not even source their "some-of-the-worst-air-pollution" assertions, and everyone "knows" that more traffic means more air pollution. This misinformation has quietly become an unquestioned part of the zeitgeist—much to the detriment of realistic understanding of and public dialogue on air pollution issues.

6

How Many Americans Live in Areas That Violate Federal Air Pollution Standards? Far Fewer Than You Think

The number of people living in areas that violate the federal 8-hour ozone standard declined from 86 million to 35 million between 2002 and 2005. For $PM_{2.5}$, the decline was from 65 million to 43 million. Not only are these numbers only a fraction of what regulators and activists claim; this remarkable improvement has gone almost completely unnoticed.

One reason for this failure to recognize progress is the way in which EPA and others go about determining how many people breathe air that exceeds federal pollution limits. We showed in chapter 5 that many U.S. counties have some areas that violate air pollution standards and others that comply with them. EPA considers an entire county in violation of a given standard if even one monitor violates it. This can make sense for air quality *regulation*, because air pollution is often a regional problem, with emissions in one area affecting pollution levels in another. Furthermore, because state and local governments are ultimately on the hook for ensuring compliance with federal requirements, county-based nonattainment designations can harmonize Clean Air Act responsibilities with preexisting government entities.

Using this same system for estimating or communicating public health risks makes no sense, however, because only air pollution where you live, work, and breathe can affect your health. Air quality ten or twenty miles away from you is irrelevant.

Figure 6-1 on the following page shows that EPA and ALA overestimate air pollution exposure in many counties. The figure displays the percentage of monitoring sites in each county that violated the 8-hour

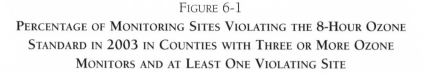

FIGURE 6-1

PERCENTAGE OF MONITORING SITES VIOLATING THE 8-HOUR OZONE
STANDARD IN 2003 IN COUNTIES WITH THREE OR MORE OZONE
MONITORS AND AT LEAST ONE VIOLATING SITE

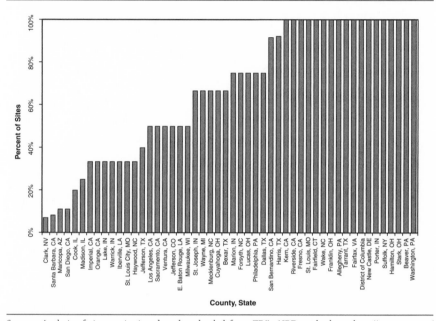

SOURCE: Analysis of site monitoring data downloaded from EPA's AIRData database, http://www.epa.gov/
air/data/geosel.html (accessed September 29, 2006).

ozone standard as of the end of 2003 and includes all counties with at
least three monitoring sites. Note that most counties had at least one site
that attained the 8-hour standard. Some of the most populous counties in
the country—for example, Cook (Chicago), Maricopa (Phoenix), and San
Diego—attained the 8-hour standard at all but a few sites in 2003. ALA
gives an entire county a failing grade for air quality even if only a small
portion violates EPA's 8-hour ozone standard.

ALA even gives failing grades to some counties when *all* of their pollution
monitors comply with federal standards. All the sites in Cook and Maricopa
counties complied with the 8-hour ozone standard as of the end of 2004 and
stayed in compliance for 2005. But *State of the Air: 2006*, which is based on
data through 2004, failed both counties for ozone anyway.[1]

Based on these counts derived from pollution monitoring data for 2002–4, ALA claimed in 2006 that 141 million Americans live in areas that violate the 8-hour ozone standard.[2] The actual number for 2002-4, however, was about forty-eight million Americans—about one-third ALA's claim.[3] Based on data for 2003–5, the most recent available when ALA issued *State of the Air: 2006*, the real number was twenty-nine million people, or about one-fifth ALA's claim.

EPA similarly overestimates ozone exposure, claiming 159 million Americans live in areas that violate the 8-hour standard.[4] EPA's latest estimate is for data through 2003. The real value for that year is seventy million, or less than half of EPA's claim.

ALA's and EPA's overestimations are even greater for $PM_{2.5}$. About thirty million Americans lived in areas that violated federal $PM_{2.5}$ standards as of the end of 2004 and about thirty-three million as of the end of 2005.[5] ALA claimed sixty-four million and EPA claimed eighty-eight million, based on data through 2004.

ALA's and EPA's overcounts result mainly from the fact that they include clean areas of counties as having air that violates federal ozone or $PM_{2.5}$ standards. ALA also uses tougher standards than the actual federal standards, while EPA includes some counties in its tally that fully attain the standard.[6]

Not only do regulators and activists continue to overestimate the number of people breathing air that violates federal pollution standards; their analyses simply ignore the recent unprecedented decline in U.S. air pollution levels since 2002. From 2002 to 2005, the number of Americans living in areas that violated the 8-hour ozone standard or the annual $PM_{2.5}$ standard declined 59 and 34 percent, respectively.[7] The news media, which rely on activists and regulators for their air pollution information, have been mum on this as well.

Why has this good news gone unreported? The countywide method of tallying population in nonattainment areas used by regulators and activists masks the apparent reduction. For example, if a county goes from 90 percent of the population living in violating areas down to 10 percent, EPA and ALA still count everyone in the county as breathing polluted air.

Who Really Lives in Areas That Violate
Federal Pollution Standards?

EPA's and ALA's method of counting how many Americans breathe polluted air is inaccurate, but it does have the virtue of being simple to implement, because county boundaries and populations are relatively easy to determine. A more sophisticated approach is necessary to make realistic estimates. Here we provide a brief summary of our approach. The appendix to this chapter on page 213 provides more detail on the methodology.[8]

Even within a city, most people live miles from the nearest air pollution monitor. Thus, to estimate pollution levels in areas between monitors, we use a method called interpolation. This just means that we estimate the pollution levels at each census tract in a metropolitan area by averaging the levels at the nearest monitor in each of eight directions (north, northeast, east, southeast, and so forth) from the center of the census tract—a method known as an octal search. Our averaging technique is an inverse-distance-weighted average. This means that in estimating pollution levels at a given location, we give more weight to nearby pollution monitors than to those that are farther away.

Our interpolation method works well in areas that have at least a few pollution monitors. More than 70 percent of Americans live in counties monitored for ozone and $PM_{2.5}$ (often, but not always, the same counties). However, most counties don't have any pollution monitors. These tend to be outside metropolitan statistical areas (MSAs)—that is, their population tends to be low, and their residents tend to live in small towns or rural areas. We assume that the violation rate in unmonitored counties in a given state is the same as that in the non-MSA counties in the state that do have monitors. So, for example, if 15 percent of monitoring locations in non-MSA counties in a given state violate the 8-hour ozone standard, we assume the same violation rate for unmonitored counties. We then multiply the population in these unmonitored counties by 0.15 to estimate the number of people living in areas that violate the 8-hour standard.

Figure 6-2 summarizes the numbers and percentages of Americans living in areas that violated federal 8-hour and 1-hour ozone and annual $PM_{2.5}$ standards during 2002–5. The right side of the chart also estimates

FIGURE 6-2

NUMBER AND PERCENTAGE OF AMERICANS LIVING IN AREAS THAT VIOLATED FEDERAL OZONE AND PM$_{2.5}$ STANDARDS, 2002–5

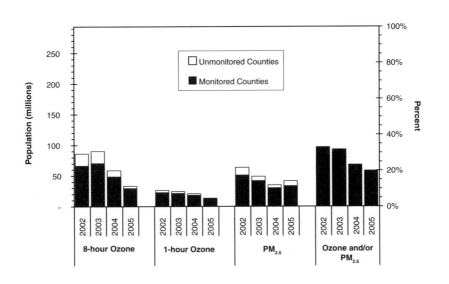

SOURCES: Air pollution data were downloaded from EPA's Air Quality System (AQS) database at http://www.epa.gov/ttn/airs/airsaqs/detaildata/downloadaqsdata.htm and http://www.epa.gov/ttn/airs/airsaqs/archived%20data/downloadaqsdata-o.htm (accessed November 27, 2006). Population by census tract: U.S. Census Bureau, Census 2000 Census Tract Population and Population Centers, October 2002, http://www.census.gov/geo/www/cenpop/tract/tract_pop.txt (accessed December 7, 2006).

NOTES: "Ozone and/or PM$_{2.5}$" includes only counties monitored for at least one of the two pollutants.

the number of people living in areas that violated standards for either or both of the pollutants. We don't break out results separately for the 24-hour PM$_{2.5}$ standard because hardly any monitors violated the 24-hour standard, and all of these also violated the annual standard.

The chart includes estimates for counties with at least one monitoring site for the given pollutant, and also rougher estimates for unmonitored counties based on the method described above.[9] Among counties with pollution monitors, we estimate that 34 million, or 11 percent, of Americans lived in areas that violated the 8-hour ozone standard as of the end of 2005. This included 29 million people in monitored counties and 4 million in unmonitored counties. For $PM_{2.5}$, about 41 million Americans, or 14 per-cent, lived in areas that violated federal standards—33 million in monitored counties and 8 million in unmonitored counties. Among counties moni-tored for at least one pollutant, 58 million, or 20 percent, of Americans lived in areas that violated either ozone or $PM_{2.5}$ standards, or both.

Note the large decrease in the number and fraction of Americans living in areas that violated federal pollution standards—a 62 percent decline for 8-hour ozone, a 35 percent decline for $PM_{2.5}$, and a 42 percent decline for the combined pollutants.

Pollution monitoring data for 2006 became available shortly before this book went to press. As shown in chapters 1 and 2, both ozone and $PM_{2.5}$ were lower in 2006 than in 2003. Thus, the number of Americans living in violating areas is now even lower than for 2005.[10]

Figure 6-3 compares our population estimates, which are tied to actual pollution levels where people live, with EPA's and ALA's claims. Because EPA's and ALA's estimates are only for counties with pollution monitors, we include only monitored counties for our estimates as well. As the chart shows, both organizations overstate by a large margin Americans' exposure to air pollution.

In September 2006, EPA adopted a new 24-hour $PM_{2.5}$ standard of 35 $\mu g/m^3$.[11] Although virtually the entire nation now attains the current 65 $\mu g/m^3$ standard, the new standard is much more stringent. Based on data for 2003–5, the $PM_{2.5}$ violation rate would rise from 16 percent of monitoring sites to 26 percent. As shown in figure 6-4 on page 116, this would increase the number of people living in violating areas by more than a factor of two, from 33 million to 74 million (monitored counties only).

ALA's and EPA's overestimates would make little difference if they were confined to a few technical reports. But they are repeated over and over again in news stories, op-eds, fact sheets, conference presentations, and professional publications.[12] Like inaccurate claims about air pollution lev-els discussed in the last chapter, these claims about pollution exposure have attained the status of "common knowledge"—something that "everyone knows," but that is not true.

In fact, slightly more than 20 percent of all Americans, rather than more than half, live in areas that actually violate federal standards for ozone, $PM_{2.5}$, or both. For each pollutant individually, the fraction is lower still—

FIGURE 6-3

COMPARISON OF ACTUAL NUMBER OF AMERICANS LIVING IN AREAS
VIOLATING FEDERAL POLLUTION STANDARDS WITH EPA AND ALA CLAIMS

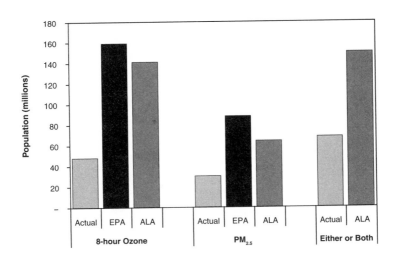

SOURCES: For our estimates, air pollution data were downloaded from EPA's Air Quality System (AQS) data-base, http://www.epa.gov/ttn/airs/airsaqs/detaildata/downloadaqsdata.htm and http://www.epa.gov/ttn/airs/airsaqs/archived%20data/downloadaqsdata-o.htm (accessed November 27, 2006). Population by census tract: U.S. Census Bureau, Census 2000 Census Tract Population and Population Centers, October 2002, http://www.census.gov/geo/www/cenpop/tract/tract_pop.txt (accessed December 7, 2006). EPA's claim: U.S. Environmental Protection Agency, *Latest Findings on National Air Quality, 2002 Status and Trends*, September 2003, http://www.epa.gov/air/airtrends/aqtrnd02/2002_airtrends_final.pdf (accessed November 27, 2006); ALA's claim: American Lung Association, *State of the Air: 2006*, http://lungaction.org/reports/stateoftheair2006.html (accessed September 29, 2006)..

NOTES: All estimates are based on pollution monitoring data for 2002–4. EPA has not provided an "either or both" estimate for this period but did estimate 146 million people living in areas that violated at least one pollution standard in 2002.

about 11 percent for ozone and 14 percent for PM$_{2.5}$. And because average pollution levels have been dropping, even areas that still violate a federal standard do so by smaller margins than in the past. Nevertheless, we do estimate that tens of millions of Americans live in areas that violate one or more standards. We address the health implications of this in the next chapter.

FIGURE 6-4

NUMBER AND PERCENTAGE OF AMERICANS LIVING IN AREAS THAT
VIOLATE CURRENT AND PROPOSED FEDERAL PM$_{2.5}$ STANDARDS, BASED ON
MONITORING DATA FOR 2003–5

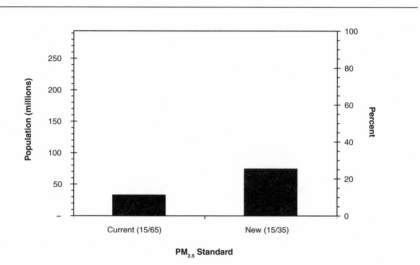

SOURCES: PM$_{2.5}$ data were downloaded from EPA's Air Quality System (AQS) database, http://www.epa.gov/
ttn/airs/airsaqs/detaildata/downloadaqsdata.htm and http://www.epa.gov/ttn/airs/airsaqs/archived%20data/
downloadaqsdata-o.htm (accessed November 27, 2006). Population by census tract: U.S. Census Bureau,
Census 2000 Census Tract Population and Population Centers, October 2002, http://www.census.gov/geo/
www/cenpop/tract/tract_pop.txt (accessed December 7, 2006).

NOTES: The annual standard will remain at its current level of 15 µg/m^3. The new 24-hour standard is
35 µg/m^3. The phrase "15/65" is shorthand for the current PM$_{2.5}$ standards, and "15/35" is shorthand for
the new standards EPA recently adopted, which will be enforced starting in 2010.

7

Air Pollution and Health

Researchers from the University of Southern California (USC) caused a stir in February 2002 when they reported that children who played three or more team sports were more than three times as likely to develop asthma if they lived in a high-ozone area than if they lived in a low-ozone area.[1]

Dozens of newspapers around the country covered a press conference on the results held by the USC researchers and officials from the California Air Resources Board (CARB), which funded the study.[2] Many journalists quoted medical experts or regulators who claimed the results were applicable to cities all over the nation and demonstrated the need for tougher air quality regulations to protect children. The study is now frequently cited by journalists, activists, and health researchers to support claims that air pollution causes people to develop asthma.

But unmentioned at the press conference was that, while higher ozone was associated with a greater risk of developing asthma for children who played three or more team sports (8 percent of children in the study), it was also associated with a 30 percent lower risk of asthma in the full sample of children in the study.[3] Other air pollutants, such as nitrogen dioxide and particulate matter, were also associated with a *lower* risk of developing asthma.

Also contrary to what the regulators and health scientists claimed, the results of the study didn't apply anywhere else in the nation. The ozone levels associated with an increased risk of asthma in child athletes were unique to the few high-ozone Southern California areas where the study was performed. In fact, by the time the findings were released in 2002, the results didn't even apply in those areas. The study was based on ozone levels during 1994–97, but 8-hour exceedances had declined 55 percent in the interim.

In short, the asthma results from the California study were just the opposite of what health scientists and regulators portrayed them to be.

117

Unfortunately, the many journalists who covered the study's release reported only what they were told, rather than what the study actually found.

In a nationwide survey conducted in May 2004, 85 percent of Americans rated air pollution as a "very serious" or "somewhat serious" problem, with similar results for state surveys.[4] These fears are not surprising, because most popular information about air pollution is alarming. Activist groups issue a steady stream of reports with scary titles such as *Danger in the Air; Death, Disease and Dirty Power; Highway Health Hazards; Plagued by Pollution;* and *Children at Risk.*[5] The American Lung Association's annual *State of the Air* reports claim most people living in American cities are "at risk" from air pollution. Regulators declare "code orange" and "code red" alerts on days when air pollution is predicted to exceed federal health standards.

Health researchers often issue alarming summaries of their research as well. Recent press-release headlines from health research institutes include "Smog May Cause Lifelong Lung Deficits"; "Link Strengthened Between Lung Cancer, Heart Deaths and Tiny Particles of Soot"; "USC Study Shows Air Pollution May Trigger Asthma in Young Athletes"; and "Traffic Exhaust Poisons Home Air."[6] News stories on air pollution often feature scary headlines such as "Air Pollution's Threat Proving Worse Than Believed"; "Don't Breathe Deeply"; "Study Finds Smog Raises Death Rate"; "State's Air is Among Nation's Most Toxic"; and "Asthma Risk for Children Soars with High Ozone Levels."[7]

Headlines like these might be warranted if they accurately reflected the weight of the scientific evidence. But they do not.

For example, in 2005, to justify a tougher ozone standard adopted for California, the California Air Resources Board prepared a detailed report summarizing research results on ozone's health effects.[8] In this nearly 2,000-page report analyzing hundreds of studies, CARB failed to mention a CARB-sponsored study reporting that *higher* ozone was associated with a *lower* rate of hospital visits in California's Central Valley.[9] EPA also failed to mention this study in its latest review of the federal ozone standard.[10]

The USC asthma study was part of a larger CARB-sponsored effort known as the Children's Health Study (CHS). In another CHS report, USC researchers assessed children's lung development from ages ten to eighteen.[11] In this case, ozone had no effect on lung development, but children who grew up in areas with the highest $PM_{2.5}$ levels in the country had about 2 percent less lung capacity than those in areas with background PM$_{2.5}$ levels. Based on these results, a USC/National Institutes of Health (NIH) press release sounded the alarm: "Teens in Smoggy Areas at High Risk for Starting Adulthood with Serious Lung Deficits."[12]

The press release didn't mention that researchers' definition of "serious lung deficits" was a 2 percent reduction. Indeed, even the journal article on the study didn't quantify the decrease explicitly, which can be calculated by combining numbers found in three different sections of the article. NIH's hyperbole aside, even the 2 percent decrease was misleading. It applied only in an area of Southern California with uniquely high PM$_{2.5}$ levels— more than two times the federal standard for annual-average PM$_{2.5}$ at the time the study was performed. PM$_{2.5}$ levels had also dropped more than 30 percent in the high-PM areas between the period when the study was performed (1994–2000) and the year in which it was published (2004). Furthermore, the children in the study were already ten years old in 1994 when the study began. They'd experienced far higher PM$_{2.5}$ levels from 1984 to 1994. These earlier PM$_{2.5}$ levels are even less relevant to the much lower national PM$_{2.5}$ levels of the late 1990s and early 2000s.

EPA's annual PM$_{2.5}$ standard is based largely on the results of a study known as the American Cancer Society (ACS) cohort study. The ACS study reported an association between long-term exposure to higher PM$_{2.5}$ levels and an increased risk of death.[13] But a reanalysis of this study showed that the association disappeared when researchers used a more complete statistical model with more extensive controls for confounding factors.[14]

Later analysis also revealed biologically implausible anomalies, suggesting that the study is turning up chance correlations rather than real causal connections. For example, PM$_{2.5}$ appeared to kill men but not women; those with no more than a high school education but not those with some college; and the moderately active but not the very active or the sedentary.[15]

Other studies question the health effects of air pollution at current exposure levels. One such study of fifty thousand veterans with high blood pressure found no association between long-term PM$_{2.5}$ exposure and risk of death, even though these men should have been more susceptible to harm from PM$_{2.5}$ than the average person.[16]

Coal-fired power plants have been one of environmentalists' premier targets during the last several years. In reports such as *Danger in the Air, Death, Disease and Dirty Power, Children at Risk*, and many others, environmental groups claim that particulate pollution from power plants is killing thousands of Americans each year.[17] There's just one problem: ammonium sulfate is the main form of PM from power plants, and it is not toxic, even at many times the highest levels ever found in ambient air.[18] In fact, ammonium sulfate is used as an inert control—that is, a compound not expected to have any health effects—in studies of the health effects of acidic aerosols.[19] Furthermore, asthma medications to open airways, such as albuterol, are delivered in the form of sulfates. A small portion of particulates of power plants are in the form of nitrates (formed from NOx emissions), and these too are not toxic.[20] In other words, the campaign against coal-fired power plants is based on a fundamentally false premise.

The American Lung Association's Web site includes an area called "Medical Journal Watch," which summarizes hundreds of air pollution health studies.[21] But the site omits studies that do not report any harm from air pollution, such as the veterans study of $PM_{2.5}$ and mortality mentioned above.[22] And for studies that report mixed results, ALA usually mentions only the harmful effects. For example, ALA discusses the CHS lung-growth study cited above, but does not mention that ozone had no effect on lung growth.

Through exaggeration, selective inclusion and omission of information, and lack of context, regulators, activists, and even many health experts create the appearance that harm from air pollution is much greater and more certain than suggested by the underlying scientific evidence.

The public's interest is in an accurate portrayal of air pollution risks. Incorrect portrayals foment fear out of all proportion to the negligible risks posed by air pollution at contemporary low levels. It distracts attention from larger and more certain risks. And it creates a political environment in which regulators and politicians can impose costly regulations that provide tiny or zero benefits in exchange for enormous costs.

We all pay in the form of higher prices for useful goods and services, lower wages, lower returns on investments, and reduced choices. Because we live in a world of many needs and desires but scarce resources, when we choose more air pollution regulation, we are also implicitly choosing less of other things we care about, like health care, education, housing, food, and vacations. By choosing further pollution reductions for air that is already safe, we have made these tradeoffs poorly, to the detriment of our overall health and welfare.

As we show in this chapter, air pollution affects far fewer people, far less often, and with far less severity than Americans have been led to believe.

How Do Scientists Assess Air Pollution's Health Effects?

In general there are two ways to study the health effects of air pollution (or anything else that might be associated with health risks or improvements): randomized trials or nonrandom "observational" studies. Randomized trials may be familiar in the form of drug trials. People in a study of a potential new drug are randomly assigned to either a "treatment" group or a "control" group. The treatment group receives the drug being studied, while the control group receives a placebo—something that looks the same as the treatment but is really just a sugar pill or some other inactive substance.

Through random assignment, researchers ensure that the treatment and control groups start out exactly the same. Then, any differences between the two at the end of the trial can be confidently presumed to have been caused by the treatment, rather than by other factors. Randomized trials are often referred to as "experimental" studies, because they are like a laboratory experiment in which researchers remove all potential sources of bias and nonrandom variation in order to isolate the unique properties and behavior of the substance being studied.

Random assignment is important because many factors, both known and unknown, can affect health. For example, if people in the control group were more likely to be smokers, or to eat fattier foods, then researchers might mistakenly attribute benefits to a drug that were really "caused" by the treatment group having better health habits than the control group. Or, a drug could have benefits that might be obscured if the treatment group has worse health habits than the control group. Random assignment eliminates these sources of bias by ensuring that the treatment and control groups are the same in all important observable ways.

It is sometimes possible to perform randomized trials on the health effects of air pollution. For example, researchers can expose randomly selected groups of animals to dirty or clean air in a laboratory to see if the pollution causes any detrimental effects under these controlled conditions. Mild, temporary health effects can be studied with small groups of human volunteers. For example, people can be exposed to ozone or $PM_{2.5}$ in a laboratory chamber, perhaps while exercising. Their lung function can be measured at various times after exposure to various levels of pollution, including clean air, to see if it, say, reduces lung capacity or increases susceptibility to allergic reactions.

Unfortunately, in most cases, it is not possible to perform randomized trials to assess human health risks of air pollution, because they are either too expensive or unethical. For example, we can't, of course, do a randomized trial of whether air pollution kills people. For one thing, people can't be randomly assigned to live for months or years in areas of varying air pollution levels. But even if we could imagine doing such a study, this still wouldn't tell us whether air pollution kills. We might find higher mortality associated with living in areas with higher pollution, but we wouldn't know if the pollution is actually the cause, because these areas might differ from low-pollution cities in other ways that affect health, say different weather or types and amounts of pollen. It might be these other, possibly unknown, nonpollution factors that are actually the cause of the observed higher mortality in high-pollution cities. In addition, since air pollution is a mixture of many substances, even if it were the cause of higher mortality we wouldn't know which particular substance or substances were the actual culprits.

In the absence of randomized trials, researchers turn to "observational" or "nonexperimental" studies to assess health risks. Observational studies work with nonrandomly selected subjects and nonrandomly assigned pollution exposures and then use statistical methods to try to remove the biases inherent in nonrandom data. Most epidemiological studies you read about in the newspaper—studies that assess the effects of diet or health habits on risk of cancer or heart disease, for example—are of this nonrandomized, observational sort.

In the case of air pollution, researchers might follow a large group, or "cohort," of people from several different cities and see if there is any correlation over time between pollution levels and risk of dying. This is known as a cohort study. Or they might look within a city for correlations between daily fluctuations in pollution levels and daily numbers of deaths or hospital admissions, which is known as a time series study.

Since observational studies are based on nonrandom data, they risk turning up correlations between pollution and health effects that might not be due to pollution at all. For example, if people who live in high-pollution cities are more likely to be overweight or to smoke, then cohort studies might attribute health effects to air pollution that were actually caused by poor health habits. This problem is called "confounding." Confounding occurs when some third factor is correlated with both health status and with pollution levels. Air pollutants can also cause confounding among themselves, because levels of various pollutants can be correlated with each other.

Time series studies can also suffer from confounding, in this case due to weather. For example, higher temperatures are associated with higher ozone levels. But heat stress increases people's risk of death independently of air pollution levels, so inadequately accounting for weather could cause one to conclude that ozone increases mortality when, in fact, weather was the culprit. Long-term trends in a population's health status can also cause confounding. Many unmeasured factors, such as demographic changes or changes in health care over time, affect mortality rates and show up as trends in mortality over time. Seasonal trends in health risks can also be correlated with pollution levels.

The output of an observational epidemiology study is a correlation between some factor, say air pollution levels or dietary fat, and a health outcome, such as death, atherosclerosis, or an asthma attack. But unlike laboratory studies and clinical trials, "gold standard" methodologies that produce direct evidence for cause-effect relationships through random selection and assignment of subjects, the evidence from observational studies is indirect.

Researchers try to control for confounding in their statistical models by adding, say, income and smoking status as control variables in cohort studies, or by adding weather and time-trend variables in time series studies. The goal is to remove statistically all nonpollution sources of variability—in effect, to turn nonrandom data into the equivalent of treatment and control groups. The assumption is that after controls are added for known or expected confounders, any residual correlation between air pollution levels and health outcomes represents a genuine causal linkage.

Several lines of evidence indicate that this assumption is mistaken and that observational studies often give false indications of risk. Experience has shown that adequately controlling for all important confounders is exceedingly difficult, if not impossible. The evidence for this difficulty comes not from air pollution studies but from traditional health studies. For example, based on observational studies of hormone-replacement therapy (HRT), medical researchers concluded that not being on HRT increased a woman's risk of developing heart disease by a factor of two.[23] An influential analysis of these studies, published in 1991, helped make HRT one of the most frequently prescribed therapies in the United States.[24] But more recently, randomized trials, which eliminate the possibility of confounding by unobserved factors that affect health, indicated that HRT doesn't reduce heart disease risk and might even increase risk.[25]

In fact, most health claims based on observational studies are turning out to be false when tested in randomized trials.[26] For example, a study of nearly forty-nine thousand women reported that following a low-fat diet for eight years did not reduce women's risk of heart disease, breast cancer, or colorectal cancer, as had been suggested by observational epidemiology studies.[27] Another randomized trial recently showed that calcium and vitamin D supplements do not reduce women's risk of osteoporosis.[28] Observational studies suggested that beta-carotene (vitamin A) supplements reduce people's risk of dying from heart disease by about 30 percent. But randomized trials have reported a 12 percent *increase* in risk of death from beta-carotene supplements.[29]

Why do observational studies give spurious results? Two serious and pervasive problems are publication bias and data mining. Publication bias refers to the tendency of researchers to seek publication of, and for scientific journals to accept, mainly those studies that find an "expected" effect.[30] Studies that find no effect ("negative" studies) or an effect contrary to the researcher's expectations or to the prevailing wisdom in a given field are more likely to end up in a file drawer. As a result, the real effect of any particular diet, medical intervention, air pollutant, for example, is smaller, often much smaller, than the sum of studies in the scientific literature would naively lead one to believe.

Data mining refers to the tendency of observational studies to become statistical fishing expeditions that turn up chance correlations, rather than real causal relationships. Think of the statistical models that researchers use to control for bias in epidemiological studies as having a lot of "dials" that researchers can turn to "tune" the results. Researchers tend to turn these dials in ways that maximize the effects they expect or hope to find and that are more likely to get the research accepted in a prestigious journal. As a recent review of air pollution epidemiology cautioned:

> Publication bias arises because there are more rewards for publishing positive or at least statistically significant findings. It is a common if not universal problem in our research culture. In the case of time-series studies using routine data there are particular reasons why publication bias might occur. One is that the data are relatively cheap to obtain and analyse, so that there may be less determination to publish "uninteresting" findings. The other is that each study can generate a large number of results for various outcomes, pollutants and lags and there is quite possibly bias in the process of choosing amongst them for inclusion in a paper. In the field of air pollution epidemiology, the question of publication bias has only recently begun to be formally addressed.[31]

Another review article observed that

> estimation of very weak associations in the presence of measurement error and strong confounding is inherently challenging. In this situation, prudent epidemiologists should recognize that residual bias can dominate their results. Because the possible mechanisms of action and their latencies are uncertain, the biologically correct models are unknown. This model selection problem is exacerbated by the common practice of screening multiple analyses and then selectively reporting only a few important results.[32]

Spurious risk estimates due to data mining and publication bias are a general and pervasive problem in observational epidemiology. Indeed, many health researchers are beginning to question the validity of much of the epidemiologic literature.[33] Several new journals have been created specifically to combat the problem of false health claims based on observational studies.[34] For example, the *Journal of Negative Results in Biomedicine* and the *Journal of Negative Observations in Genetic Oncology* seek to combat publication bias by publishing only "negative" studies—that is, studies that look for but fail to find a particular health effect. *The Journal of Spurious Correlations* seeks to combat data mining in sociological studies. More such journals are on the drawing board. The American Association for the Advancement of Science, which publishes the prestigious journal *Science*, held a three-hour panel discussion on the problem at its annual meeting in February 2007 titled "Mixed Health Messages: Observational Versus Randomized Trials."[35]

Unfortunately, this recognition of the limitations and weaknesses of observational studies has not filtered into the relatively insular air pollution epidemiology community. The claim that current, historically low air pollution levels can cause serious harm is based on the results of observational studies, rather than randomized trials. Regulators and air pollution epidemiologists point to the thousands of observational studies that have reported a positive association between low-level air pollution and risk of death as proof that the effects are real. But this ignores the fact that implementing an invalid methodology over and over again doesn't improve its validity.

Laboratory studies with randomized groups of animals could, in principle, provide more reliable support for the claim that air pollution at ambient levels can kill. But, as we will see below, such studies have instead shown that animals don't die even when exposed to much larger amounts of air pollution than ever occur in the real world.

In the remainder of this chapter, we demonstrate that the health rationale for continued tightening of air pollution standards rests on a surprisingly weak foundation.

Does Air Pollution Kill?

There is no question that high levels of air pollution can kill. About four thousand Londoners died during the infamous five-day "London Fog" episode of December 1952, when soot and sulfur dioxide soared to levels tens of times greater than the highest levels experienced in developed countries today, and visibility dropped to less than twenty feet.[36]

The question for policy now is whether current, relatively low levels of air pollution are also deadly. EPA created its $PM_{2.5}$ standards based on observational cohort and time series studies that reported exposure even to low levels of $PM_{2.5}$ can kill. Based on the results of these studies, scientists, activists, and regulators have concluded that PM kills tens of thousands of Americans each year.[37]

The deaths from the London Fog were obvious, because the daily number of deaths soared to many times above the "background" level. However, if recent observational studies are correct, current PM levels are increasing the chance of dying by at most a few tenths of a percent to a few percent above the baseline level. About four million people die each year in the United States. Thus, if PM were really increasing people's risk of death by, say, 2 percent, it would mean eighty thousand additional premature deaths each year. Regulators, activists, and many health experts have portrayed studies of air pollution and mortality as providing definitive proof that particulate matter, and more recently ozone, kill even at contemporary low levels. But this claim is based solely on the results of observational epidemiology studies, with little or no support from more reliable study designs. Furthermore, these claims ignore contrary evidence, both in the form of weaknesses in the studies favored by proponents of an air pollution–mortality link, and from other studies not cited in popular portrayals of air pollution's health effects.[38]

EPA set its annual $PM_{2.5}$ standard based mainly on two observational cohort studies of long-term PM exposure and mortality: an American Cancer Society study and the Harvard Six Cities (HSC) study.[39] The studies compared death rates over time for people living in different cities with PM levels in those cities. Both studies reported that the chance of dying over the period of years studied—seven for the ACS and fourteen to sixteen for the HSC—increased by several percent for each 10 µg/m³ increase in long-term $PM_{2.5}$ levels. The studies were also the subject of a detailed reanalysis by the Health Effects Institute (HEI), which ostensibly confirmed their results, as did a follow-up report on the ACS.[40] Despite the public portrayals of these studies, their actual results suggest that PM at current levels isn't killing people.

For example, the follow-up ACS study and the HEI reanalysis reported that $PM_{2.5}$ kills those with no more than a high school degree, but not those with at least some college; men but not women; and the moderately active but not the very active or sedentary. These odd variations in PM's ostensible effects don't seem biologically plausible and suggest that the apparent effect of $PM_{2.5}$ is actually spurious, resulting from failure to control adequately for confounding factors unrelated to air pollution.

The original ACS study covered the period 1982–89 and reported a 6.9 percent increase in risk of death for each 10 Ìg/m3 increase in long-term $PM_{2.5}$ levels. The follow-up report covered the period 1982–98 and reported a 4 percent increase in the risk of death. If PM were really the causal factor, this means the increased risk from $PM_{2.5}$ declined to 1.1 percent for the 1990–98 period, for a 73 percent decline in the size of the PM–mortality relationship between the 1980s and the 1990s. This smaller effect size would also be statistically insignificant—that is, indistinguishable from zero.

The ACS researchers did not point out this key result of their analysis. They reported only results for the initial follow-up period (1982–89) and then for the entire follow-up period (1982–98). However, the results for the second follow-up period—1990–98—can be inferred from the data presented in the two reports on the study.[41] Other studies suggest a similar decline over time in apparent $PM_{2.5}$ effects.[42] The fact that the apparent effect of any given level of $PM_{2.5}$ declined by more than half between the 1980s and '90s would seem to have great relevance for policy, but this has gone largely unremarked.

Reanalysis of the ACS data has also shown that considering additional factors in the statistical analysis can make the apparent $PM_{2.5}$ effect disappear. For example, when migration rates into and out of cities were added to the statistical model relating $PM_{2.5}$ and premature death, the apparent effect of $PM_{2.5}$ declined by two-thirds and became statistically insignificant.[43]

Cities that lost population during the 1980s—Midwest "rust belt" cities that were in economic decline—also had higher average $PM_{2.5}$ levels. The hypothesis is that people who work and have the wherewithal to migrate are healthier than the average person. Thus, the apparent effect of $PM_{2.5}$ could actually have resulted from healthier people moving away from areas of the country that were in economic decline, rather than from a change in any individuals' health status due to PM exposure. This example demonstrates the limitations of observational studies. Without a randomized assignment to high and low levels of air pollution, there are almost certain to be unobserved differences between groups that confound the association of air pollution and health, creating spurious results.

The Harvard Six Cities study suffers from similar problems. It, too, found no association between $PM_{2.5}$ and mortality for people with more than a high school education. The HSC study also reported a statistically significant *decrease* in mortality due to respiratory causes in areas with higher $PM_{2.5}$ levels.[44]

Three other epidemiology studies have concluded that $PM_{2.5}$ is not associated with increased mortality. One followed fifty thousand male veterans with high blood pressure—a group that should have been more susceptible than the average person to any negative effects of air pollution—over a period of twenty-one years and found that higher $PM_{2.5}$ was not associated with increased risk of death.[45] Another assessed pollution and mortality at the county level and concluded that if $PM_{2.5}$ increased risk of death, the threshold was at least $23\mu g/m^3$, which is higher than current $PM_{2.5}$ levels in more than 99 percent of the country and 50 percent greater than EPA's annual $PM_{2.5}$ standard.[46] More recently, another cohort study followed fifty thousand elderly Californians from 1973 to 2002 and reported that $PM_{2.5}$ was associated with a small increase in risk of death during the 1970s, but not from the 1980s onward.[47]

The American Lung Association's Medical Journal Watch Web site discusses hundreds of studies of air pollution and mortality. But the site does not mention any of the contrary evidence above. And while ALA does mention the HEI reanalysis of the ACS and HSC studies, it claims HEI "validated" the original results.[48] Indeed, HEI did validate the results of the original papers in the sense that its investigators were able to reproduce their results. But HEI also performed a range of sensitivity analyses which showed that better control for confounding diminished or eliminated the apparent harm from $PM_{2.5}$. ALA did not report these refutations of the $PM_{2.5}$-mortality link.

Cohort studies are expensive to perform, so there are only a few of them. However, time series studies require only daily data on pollution levels and weather, and counts of daily nonaccidental deaths in a given city.

Thus, hundreds of studies have assessed the correlation between daily fluctuations in air pollution levels and risk of death. The conventional wisdom based on these studies is that typical fluctuations in daily PM levels are associated with an increase of up to a few percent in the risk of premature death.[49] More recently, a number of researchers have reported that ozone is also associated with increases in daily mortality.[50] However, these studies suffer from their own array of problems that invalidate their results.

As detrimental as data-mining and publication bias are to cohort studies, the ease of performing time series studies makes these problems all the more acute. There are many ways to slice the data, and the tendency is to choose selectively and publish only results from those statistical models that give the largest apparent effects. It is worth stressing once again the recent article by air pollution epidemiologists cautioning that the "model selection problem is exacerbated by the common practice of screening multiple analyses and then selectively reporting only a few important results."[51]

A number of studies have demonstrated the degree to which the results of air pollution time series studies depend upon researchers' judgment calls about how to model the data, rather than on any objective evidence for health effects. Conclusions about the existence of an air pollution–mortality association can vary from study to study, even when different researchers use the same datasets for the same cities.[52] Based on the variability of results given different approaches, one group of researchers concluded, "There are many possible interpretations of the data and no single conclusion is definitive."[53]

Thus, combining the results from those studies published in the research literature—a technique called meta-analysis—leads to an overestimate of air pollution's health effects. The National Morbidity, Mortality, and Air Pollution Study (NMMAPS) has demonstrated how large the effect of publication bias can be. NMMAPS does not suffer from publication bias, because it applied the same analytical methods to pollution and mortality data for ninety-five different U.S. cities and published all the results. A recent NMMAPS report on the relationship between ozone and mortality found an effect 70 percent lower than that derived from meta-analysis of single-city studies and concluded that publication bias inflates the effects estimated via meta-analyses.[54] A World Health Organization (WHO) study drew similar conclusions on the effect of publication bias.[55]

Both the WHO analysis and NMMAPS reported additional results that bolster concerns about the reliability of observational studies. One surprise from NMMAPS was that in about one-third of cities in the study, higher levels of particulate matter and ozone were associated with *lower* risks of premature death.[56] The results were also sensitive to a few outlier cities. When three cities were removed from the analysis—two with increased mortality and one with decreased mortality associated with PM—the average effect of PM across the cities in the study became statistically insignificant.[57] It isn't clear that an average pollution effect across cities even has any meaning when air pollution seems to protect people in some cities and harm them in others.

The WHO analysis also reported some biologically implausible results: There was no association between ozone and respiratory mortality, while the association of ozone with cardiovascular mortality was the same as for all-cause, nonaccidental mortality. But if ozone exerts its effects through the respiratory system, one would expect a *greater* effect on respiratory and cardiovascular mortality than on all-cause mortality. Furthermore, after adjusting for publication bias, the WHO analysis concluded that higher ozone was associated with *lower* respiratory mortality.

In July 2005, three meta-analyses of daily ozone levels and mortality appeared in the journal *Epidemiology*.[58] Each was performed by a different research group, but all were commissioned by the EPA. EPA, environmentalists, and some health researchers claimed the results provided definitive proof that ozone is killing people.

But unremarked in the press releases and media reports was that as meta-analyses, all three of these studies inherently suffered from publication bias, and two even provided evidence of that bias. A commentary accompanying the three meta-analyses pointed out the problem:

> In the absence of NMMAPS or other multisite analyses, some observers might have taken the agreement of the meta-analyses as confirmation that the meta-analytic method was reliable. However, if our observational methods are all subject to the same biases, as meta-analyses are when they are derived from the same pool of studies, the agreement criterion is testing a narrow range of assumptions.[59]

As already noted, data-mining can increase the apparent harm from air pollution when researchers report only a few results out of many different statistical models they may have screened. To address this, some studies have reported on the results of screening large numbers of plausible models relating air pollution and daily mortality. One showed that of more than 1,200 different models, more than one-third predicted that higher air pollution was associated with a lower risk of death.[60]

To address the problem of data-mining in a more comprehensive way, a few researchers have used a technique known as Bayesian model averaging (BMA). Though mathematically complicated, the technique is simple in principle: Take all possible statistical models relating air pollution and other factors, such as weather, to health outcomes; weight the models according to how well they fit the actual data; then take a weighted average of the results. This gives an average and an uncertainty range for the correlation between, say, particulate matter and death, after controlling for the effects of other factors that could affect health. These other factors include:

- Weather conditions, such as temperature, humidity, wind speed, and barometric pressure;

- Timing of effects. Current-day pollution or current-day weather might be a culprit, for instance, but delayed effects from the previous few days might be important as well;

- Interactions among variables. For example, ozone might cause death only on very hot days, or only in the presence of another pollutant;

- Other air pollutants, such as ozone, carbon monoxide, or sulfur dioxide;

- Long-term trends in mortality unrelated to air pollution, such as deaths from flu epidemics or other causes related to season, or longer-term mortality trends related to changing health habits, such as diet, exercise, or smoking.

The result is literally hundreds of potential explanatory variables and trillions of potential models. A recent study based on Bayesian model averaging concluded that the effect of air pollution on mortality is statistically indistinguishable from zero.[61] According to the researchers, "Models that elicit statements of the form 'ozone has no effect on mortality' receive the most support from the data." The researchers drew a similar conclusion for particulate matter and other pollutants. They cautioned, "A method that presents results from a single regression [that is, from a single statistical model] may lead researchers to make misleading inferences about pollution–mortality effects, thereby seriously underestimating the true uncertainty in the statistical evidence."

Considering the pitfalls of publication bias, data-mining, and uncontrolled confounding, observational studies have proved to be unreliable guides to air pollution health risks, just as such studies have proven to give false indications of health effects in other areas of health research. Emphasizing the point, a recent study showed that adding a more sophisticated and complete adjustment for the health effects of weather can cause the apparent harm from PM and ozone to disappear. When two British researchers allowed in their model for cumulative effects of heat stress over several days, as well as the additional effects of increases in direct sunshine and lower winds, both of which add to heat stress at any given temperature, the associations of air pollution and risk of death were reduced by 60–90 percent and became statistically insignificant.[62]

Given the unreliability of observational studies, the case for low-level air pollution as a cause of death might seem more plausible if researchers had shown in laboratory studies that air pollution kills animals. However, researchers have been unable to kill animals even with air pollution at levels many times greater than is ever found in ambient air.[63] A recent review of particulate matter toxicology concluded, "It remains the case that no form of ambient PM—other than viruses, bacteria, and biochemical antigens—has been shown, experimentally or clinically, to cause disease or death at concentrations remotely close to US ambient levels."[64]

This seemingly changed in December 2005, when the *Journal of the American Medical Association* (*JAMA*) published the results of a study that claimed $PM_{2.5}$ at current ambient levels is increasing Americans' risk of developing heart disease.[65] The study exposed mice to 85 $\mu g/m^3$ of $PM_{2.5}$ concentrated from ambient air for six hours a day for six months, or about one-quarter of a typical mouse lifespan.

Mice fed a high-fat diet and exposed to $PM_{2.5}$ had more than a 50 percent greater rate of atherosclerosis (as measured by arterial plaque area) and other signs of heart disease than a control group that was fed a high-fat diet but not exposed to $PM_{2.5}$. $PM_{2.5}$ was associated with greater atherosclerosis in mice on a low-fat diet as well, but the effect wasn't statistically significant.

NIH highlighted the study with a press release that began, "Test results with laboratory mice show a direct cause-and-effect link between exposure to fine particle air pollution and the development of atherosclerosis. . . . [The study] may explain why people who live in highly polluted areas have a higher risk of heart disease."[66] The study caused a minor media sensation, with both journalists and health experts claiming it provided definitive evidence that $PM_{2.5}$ is causing serious harm to human beings.[67]

Despite the enthusiastic reception, there's much less here than meets the eye. The mice used in the study were genetically engineered in ways that made them unrepresentative of even real-world mice, much less of humans. The mice were designed to lack the gene for apolipoprotein E (ApoE), a key substance for fat and cholesterol metabolism. As a result, these ApoE "knockout" mice had blood cholesterol levels five to six times greater than normal mice when fed regular rat chow. The "knockout" mice had fourteen times the cholesterol of normal mice when both were fed a high-fat diet.[68]

These are stupendous cholesterol levels. For comparison, medical authorities define "high cholesterol" as a serum cholesterol level greater than 240 milligrams per deciliter (mg/dl), which is about 20 percent greater than the average cholesterol level in American men.[69] Only one in fifty American men exceeds one and a half times the U.S. average, and only one in five hundred exceeds twice the average.[70]

The very reason for using such grossly unrealistic mice to study $PM_{2.5}$ is that $PM_{2.5}$ does not kill regular mice or other animals at concentrations relevant to real-world human exposures. For that matter, none of the super-high-cholesterol mice in the *JAMA* study died either.

NIH downplayed the vast gulf between the genetically engineered mice and normal mice, stating only that they were "genetically programmed to develop atherosclerosis at a higher-than-normal rate." This is a bit like doing a study on people who weigh five hundred pounds and referring to them merely as "overweight."

If you build a house out of cards, you would expect even a gentle breeze to knock it down. But this doesn't tell you much about the ability of a real wood-frame house to withstand a gentle breeze. Likewise, if you design an artificial mouse that can't regulate its fat or cholesterol levels, it isn't any surprise that even a minor environmental insult can cause it some health problems. But this doesn't tell you much about the effects of low-level air pollution on regular mice or on people.

Unfortunately, news articles on the study failed to provide the context that would have shown that the findings have no real-world relevance. A Nexis search turned up ten newspaper reports, of which seven didn't mention that the mice had been genetically engineered, leaving the impression that real-world $PM_{2.5}$ levels caused heart disease in normal mice.

Three other news outlets followed NIH's lead, creating the impression that the mice in the study were merely analogous to people with a higher-than-average risk of heart disease. For example, according to the *Los Angeles Times*, the mice were "bred to be susceptible to developing heart disease."[71]

NIH and the study authors also misled reporters about the relevance of the $PM_{2.5}$ doses to real-world levels. According to NIH, "The fine particle [$PM_{2.5}$] concentrations used in the study were well within the range of concentrations found in the air around major metropolitan areas." The press release also quoted one of the study's authors as saying that "the average exposure over the course of the study was 15 micrograms per cubic meter, which is typical of the particle concentrations that urban area residents would be exposed to, and well below the federal air quality standard of 65 $\mu g/m^3$ over a 24-hour period."[72]

In fact, the $PM_{2.5}$ levels in the study were nothing like real-world levels. The mice were exposed to $PM_{2.5}$ at 85 $\mu g/m^3$ for six hours in a row during five days of each week, and to filtered air the rest of the time. Over the six-month study period, this did, indeed, average out to about 15 $\mu g/m^3$, the level of the federal $PM_{2.5}$ annual standard. But in the real world, areas that average 15 $\mu g/m^3$ of $PM_{2.5}$ over a year rarely approach short-term $PM_{2.5}$ levels of 85 $\mu g/m^3$.

For example, the mice in the study spent the equivalent of 1,560 hours per year (30 hours per week x 52 weeks per year) breathing 85 $\mu g/m^3$ $PM_{2.5}$. In contrast, Modesto, California, averaged 16$\mu g/m^3$ of $PM_{2.5}$ from February 2005 to January 2006, but spent only 80 hours at 85 $\mu g/m^3$ or above.[73] Furthermore, 40 percent of those high-$PM_{2.5}$ hours occurred between 11 p.m. and 6 a.m., when most people are in bed. There were only 420 hours where Modesto exceeded even 50 $\mu g/m^3$ $PM_{2.5}$.

Even areas with the highest levels in the country have far fewer hours of high $PM_{2.5}$ than were used in the mouse study. For example, Riverside, California, has the worst $PM_{2.5}$ levels in the country. But Riverside had only 135 hours at or above 85 $\mu g/m^3$, and 1,055 hours above 50 $\mu g/m^3$ during 2005.[74]

Health effects depend not only on the average dose, but on the acute dose. For example, you could take two aspirin four times per day, or you could take eight all at once each day. Either way, your average dose is eight aspirin per day. But you're more likely to suffer ill effects if you take the aspirin all at once. The mice received an analogously unrealistic daily $PM_{2.5}$ exposure.

There's nothing wrong with the *JAMA* mouse study in principle. It shows that when you take a mouse specially designed to have unrealistically stupendous cholesterol levels, feed it a high-fat diet, and repeatedly expose it to unrealistically high acute doses of $PM_{2.5}$, the $PM_{2.5}$ increases arterial markers of heart disease. The problem arises when the study's proponents claim that this has something to do with $PM_{2.5}$ risks faced by human beings or even normal mice.

A summary of the study is now available on NIH's Web site. Its title? "Particulate Air Pollution and a High Fat Diet: A Potentially Deadly Combination."[75]

EPA attributed 90 percent of the benefits of all air pollution regulations through 1990 to PM reductions (and 8 percent to removing lead from gasoline) and 99 percent of the incremental benefits of the Clean Air Act Amendments of 1990 to PM reductions.[76] But if air pollution at current and recent levels isn't killing people, then none of these benefits are actually being realized.

Does Air Pollution Cause People to Develop Asthma?

Asthma provides a signal example of how conventional wisdom on air pollution and health can be the opposite of reality. According to the Centers for Disease Control (CDC), the prevalence of asthma in the United States rose 75 percent from 1980 to 1996, and nearly doubled for children. Prevalence may have leveled off since then.[77] Could air pollution be the cause? Not likely. Asthma prevalence rose at the same time that air pollution of all kinds declined. Figure 7-1 on the following page displays trends in asthma and various air pollutants for California, including data for ozone, carbon monoxide, nitrogen dioxide, and PM_{10}.

The trends are similar for other pollutants measured by California regulators, including $PM_{2.5}$, benzene, 1-3-butadiene, benzo(a)pyrene, perchloroethylene, xylene, lead, and many more.[78] In all cases, pollution has been declining while asthma has been rising. Data from other states tell the same story. Air pollution—at least the wide range of pollutants regulators measure and control—is not a plausible cause of rising asthma.

Despite these opposing trends, however, media and activist reports create the impression that air pollution is a major cause of rising asthma.[79] Regulators and health experts have even spun a study that found air pollution to be associated with a *lower* overall risk of developing asthma into a key piece of evidence in support of an air pollution–asthma link. Beginning in 1993, the California Air Resources Board (CARB) funded the Children's Health Study (CHS), in which researchers from the University of Southern California (USC) tracked several thousand California children living in twelve communities with widely varying air pollution levels, ranging from near-background up to the highest levels in the nation.

We began this chapter with the joint press conference in 2002 where the USC researchers and CARB managers reported that children who played three or more team sports were more than three times as likely to develop asthma if they lived in one of the six highest-ozone communities in the study, when compared with the six low-ozone communities.[80] They also claimed the study's results applied to cities all around the United States. But ironically, the CHS asthma study actually showed just the opposite. While higher ozone was associated with a greater risk of developing asthma for children who played three or more team sports (8 percent of children in the study), higher ozone was associated with a 30 percent *lower* risk of developing asthma for the full sample. While this fact was discussed in a journal article on the study, the researchers did not mention it at the press conference.[81] Higher levels of other pollutants, including nitrogen dioxide and particulate matter, were also associated with a lower asthma risk.[82] Also

FIGURE 7-1

TREND IN ASTHMA PREVALENCE VS.
TRENDS IN AIR POLLUTION IN CALIFORNIA

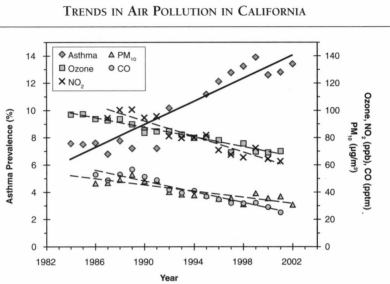

SOURCES: Asthma prevalence: California Department of Health Services, *Points of Interest—Asthma in California* (Sacramento: California Department of Health Services and the Public Health Institute, May 2003). Air pollution: California Air Resources Board, *2003 Air Pollution Data CD*. The 2003 edition of the CD is no longer available. The 2007 edition, which includes all data on the 2003 edition, plus data collected since then, is available at http://www.arb.ca.gov/aqd/aqdcd/aqdcd.htm (accessed March 21, 2007).

NOTES: The lines are linear regression lines. Ozone, CO, and NO_2 are the average of the top thirty daily readings for each year (ozone and CO peak 8-hour, NO_2 peak 1-hour) across all monitoring sites for the given pollutant. PM_{10} is the average of the annual-average PM_{10} readings for all monitoring sites. Only sites with data in every year throughout the time period for each pollutant were included in the analysis. Number of monitoring sites for each pollutant: NO_2 = 57, CO = 47, Ozone = 68, PM_{10} = 29. Pollution declined not only on average, but at almost every individual monitoring site. The start of the time period (which ranges from 1984 to 1987) for each pollutant was chosen to maximize the number of monitoring sites included, while still overlapping the period during which asthma prevalence rose. CO is listed in parts per ten million (pptm; divide by 10 in order to get parts per million) so that CO values fall within the same range as other pollutants. Ppb = parts per billion; $\mu g/m^3$ = micrograms per cubic meter.

mentioned in the journal article, but not at the press conference, was that when the researchers divided the twelve communities into three groups of four (rather than two groups of six), the association of ozone with asthma in child athletes applied only to the four communities in the highest-ozone group and not to the medium-ozone group.

The assertion that the study was relevant in other parts of the country was also false. The four high-ozone areas in the study averaged eighty-nine days per year exceeding the federal 8-hour standard and fifty-nine days per year exceeding the 1-hour standard during 1994–97, the years used to assess pollution exposure in the study.[83] No area of the United States outside a few parts of California has *ever* had ozone levels this high in a single year, much less for several years running.

In fact, by the time the study was released in February 2002, it no longer applied even in the Southern California areas where it was conducted. In the interim, 8-hour exceedances had declined 55 percent and 1-hour exceedances had declined 78 percent. This put these former "high-ozone" areas within the range of the "medium-ozone" areas, for which ozone had no effect on asthma risk.

At the press conference releasing the CHS asthma results, the chairman of the Air Resources Board claimed, "This study illustrates the need not to retreat but to continue pushing forward in our efforts to strengthen air pollution regulations."[84] But, if anything, the CHS asthma study showed that current standards already include a large safety margin. Ozone wasn't associated with a change in asthma risk in the medium-ozone areas of the study. Yet these areas averaged forty-one 8-hour exceedance days per year and seventeen 1-hour exceedances.

Inaccurate information on the Children's Health Study asthma results wasn't limited to CARB officials or USC scientists. Health experts from around the country misinterpreted the results as well. For example, on the day the study was released, a professor at the State University of New York at Stony Brook, who has since become the American Lung Association's medical director, claimed, "This is not just a Southern California problem. There are communities across the nation that have high ozone."[85] According to the *Houston Chronicle*, Houston asthma specialists said the study showed that the city should "step up its efforts to implement a state plan to reduce ozone."[86] The director of the pediatric asthma program at the University of California at Davis claimed, "Sacramento is a very high ozone area, so this [the CHS asthma study] is going to be very relevant to us."[87] Not only were all of these nominal experts wrong about whether the study was relevant to actual ozone levels in the United States; all of them completely missed the fact that ozone and other air pollutants were associated with an overall *lower* risk of developing asthma.

In a recent commentary in the *Journal of the American Medical Association*, two prominent air pollution health researchers claim, "Some evidence suggests that air pollution may have contributed to the increasing prevalence of asthma."[88] The evidence they cite is the CHS asthma study.

Environmental activists also claim or imply that air pollution is responsible for the high prevalence of asthma in the United States. For example, in its report *Clearing the Air*, the Surface Transportation Policy Project (STPP) includes a table listing the estimated amount of pollutants emitted by motor vehicles each year in dozens of metropolitan areas, along with asthma prevalence in those areas.[89] The implication is that the two are linked. But STPP never actually graphed its data to see if the emissions and asthma rates in its tables are correlated. They aren't. The left-hand graph in figure 7-2 plots STPP's estimates of asthma prevalence against its motor-vehicle emissions estimates in all metro areas for which STPP provided data. The right-hand graph plots asthma prevalence against ozone exceedances per year, also taken directly from STPP's own report. Note the lack of correlation between pollution and asthma in both graphs.

Could air pollution cause at least some people to develop asthma? The trend data suggest that asthma incidence and air pollution are unrelated. If pollution had been causing a substantial fraction of all asthma cases in the past, then the large declines in pollution over the last few decades should have resulted in a large reduction in asthma prevalence, rather than the large increase that has actually occurred.

A potential objection to this line of reasoning is that air pollution *could* be a major cause of asthma, but that some other potent asthma-causing factor(s) increased at the same time pollution declined, masking the asthma benefits from the pollution reductions. But this hypothesis doesn't wash, either. Compared to the 1980s, air pollution was far higher during the 1960s and '70s and far higher still in previous decades. Yet these "industrial doses" were not accompanied by high rates of asthma.

In addition, many developing countries have much higher air pollution than is ever found in the United States, yet they have much lower rates of asthma than the United States or other Western countries.[90] The reunification of Germany highlighted the irrelevance of air pollution for the development

FIGURE 7-2

ASTHMA PREVALENCE VS. PER CAPITA EMISSIONS FROM MOTOR VEHICLES
AND 8-HOUR OZONE EXCEEDANCES, AS REPORTED BY THE SURFACE
TRANSPORTATION POLICY PROJECT (STPP) IN *CLEARING THE AIR*

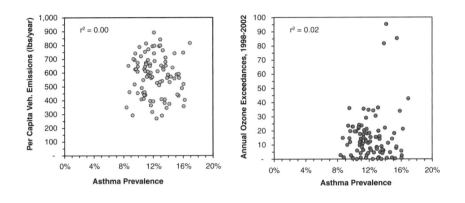

SOURCES: All data were taken directly from online appendices provided by the STPP as a supplement to its report *Clearing the Air*, http://www.transact.org/library/reports_html/Clean_Air/Appendix1.xls and http://www.transact.org/library/reports_html/Clean_Air/Appendix3.xls (both accessed November 27, 2006).

NOTES: Each point represents a Metropolitan Statistical Area (MSA). The graph includes all MSAs for which STPP reported both asthma prevalence and pollution emissions and levels. The statistic "r^2" is a measure of the relationship between two variables. An r^2 of 1 means exact correlation. An r^2 of zero means no correlation.

of asthma. Before 1991, the former East Germany had high air pollution levels and low asthma prevalence. But after reunification, at a time when the former East Germans were adopting Western lifestyles and experiencing increases in income levels, air pollution declined, but the incidence of asthma rose to levels comparable with those in the former West Germany.[91]

In 2005, the Children's Health Study produced a new wrinkle in the question of whether air pollution could cause asthma. This time the USC researchers announced that children were more likely to develop asthma if they lived closer to freeways.[92] The researchers randomly selected 208 children from the CHS and placed NO_2 monitors outside their homes for a few weeks during summer and winter 2000, and also collected data on residential distance from freeways for each child.

Living three-quarters of a mile closer to a freeway was associated with a factor of 1.9 increase in the risk of having asthma, while a 5.7 ppb increase in ambient NO_2—17 percent of the average NO_2 level in the study areas—was associated with a factor of 1.8 increase in asthma.[93] The researchers noted that, based on other studies, the evidence for NO_2 itself as a cause of asthma was weak, but suggested that NO_2 was acting as a marker for traffic-related pollution in general, especially particulates from diesel exhaust.

This suggests air pollution is having a huge effect on asthma risk. Ironically, that's what makes the study's claims so implausible. If a 17 percent change in pollution levels can cause nearly a factor of two change in the risk of developing asthma, asthma prevalence should have been many times greater back in the 1970s and early 1980s, when levels of most types of air pollution were two or three times higher than current levels. This suggests that asthma cases that the CHS researchers attribute to traffic-related air pollution are actually being caused by something else—either some unmeasured pollutant, or some other factor that is not properly or at all accounted for in their statistical analysis.

Could it be that an unmeasured pollutant in diesel exhaust has increased over the last couple of decades? This is unlikely. While we don't have information on trends in ambient levels of diesel soot, diesel-truck particulate emissions were measured several times between 1975 and 1999 in Pennsylvania tunnels. Emissions per mile declined about 83 percent over the twenty-four-year period (see chapter 4 for additional details).[94] Researchers from the University of California, Berkeley, measured particulate emissions from diesel trucks in the San Francisco Bay Area between 1997 and 2004 and recorded a 50 percent decline in diesel PM emissions.[95] According to the California Department of Transportation, total miles traveled by heavy-duty diesel trucks increased 50 percent in California between 1986 and 2001, a rate of increase far too slow to offset the benefits of cleaner trucks.[96]

The ambient $PM_{2.5}$ trend data presented in chapter 3 and the ambient benzo(a)pyrene (a diesel-related particulate) data presented in chapter 1 also indicate large declines in ambient levels of diesel particulates over time. If diesel particulates were a significant cause of asthma, we would have expected much higher asthma rates in the past—just the opposite of what we actually find. Chapters 1, 2, and 4 also show that ambient levels of gaseous pollutants have declined sharply over the last few decades.

What this asthma–traffic pollution study is missing is an analysis of trends in traffic density and air pollution over time. At any given snapshot in time, air pollution will be correlated with traffic levels and with distance from a freeway. But over time, monitoring data show air pollution has been declining both near and far from freeways, while traffic density has been increasing. Thus, a snapshot in time creates the appearance of an air pollution–asthma link even though no such causal link actually exists.

It is possible that *unmeasured* air pollutants could explain the association of asthma with living near a freeway. Tire dust or brake dust or platinum from catalytic converters might be potential culprits. However, the fact that asthma increased in all Western countries during the 1980s and '90s regardless of when these countries required catalytic converters on cars and regardless of trends in total vehicle-miles of driving suggests that this is an unlikely explanation.

There also could be socioeconomic factors that aren't captured in the statistical analyses but are correlated with both asthma risk and living near a freeway, creating the appearance of an asthma–traffic link where no causal association actually exists. We saw in the previous section how, despite researchers' best efforts, it is extraordinarily difficult to remove confounding and other statistical biases in observational epidemiology studies and that the results of observational studies are routinely overturned when their results are checked by gold-standard methods such as randomized, controlled trials. This study's small sample size might also have caused spurious results. There were only 208 children in the study, and 31 of them had asthma.

If air pollution is causing asthma, it doesn't appear that any of the wide range of substances we measure or regulate could be the culprit. The asthma–air pollution link has superficial plausibility, because asthma is a respiratory condition, but there isn't support for an air pollution link in the actual data. Nevertheless, the claim that air pollution causes asthma persists, and misleading studies like the Children's Health Study are all too common.

Medical experts often dispense erroneous information on asthma and air pollution. For example, a researcher from the Bloomberg School of Public Health at Johns Hopkins University asserted in a recent Sierra Club report that "traffic presents a unique public health threat" including "children's asthma rates occurring at epidemic proportions."[97] After the American Lung Association gave Tarrant County (Fort Worth), Texas, a failing grade for air quality in 2003, the president of a local branch of the Tarrant County Medical Society asserted, "It means we can anticipate a worsening of an already epidemic asthma problem."[98]

These and other health professionals cited earlier lend undeserved scientific authority to false air pollution–asthma scares. Now that the claim that air pollution causes asthma has become "conventional wisdom," environmentalists feel free to make ridiculous pronouncements without citing any source at all. For example, the Carolinas Clean Air Coalition's Web site claims, without attribution, that "1/3–1/2 of all asthma in North Carolina is due to air pollution."[99]

Does Air Pollution Cause Permanent Lung Damage?

Chronic exposure to high levels of air pollution can cause permanent reductions in lung function. But "high" in this case refers to ozone levels in Southern California during the 1970s and 1980s—a period when much of the region exceeded the 1-hour standard *100 to 150 days per year*. For example, a study of freshmen at Berkeley reported that students who grew up in polluted areas of Southern California scored 7–15 percent lower on lung function tests than those who grew up in the San Francisco Bay Area.[100] The average birth year for these students was 1976.

The American Lung Association cites this study to alarm people about serious long-term harm from exposure to ozone,[101] but it doesn't apply to anyone born outside Southern California, because no other area of the United States has *ever* exceeded the 1-hour ozone standard more than one hundred days per year, or even more than fifty days per year. It has been a long time since that study was even relevant in Southern California. For the last several years, even the worst four or five locations in California, and therefore the United States, have averaged only about fifteen to thirty 1-hour ozone exceedance days per year.

ALA from time to time puts out summaries of recent research on air pollution that also do more to frighten than inform. For example, under the headline, "Lung Development of Young Monkeys Drastically Changed when Exposed to Ozone Pollution," ALA summarized research at UC-Davis showing that young Rhesus monkeys who were exposed to ozone developed a disease similar to childhood asthma and concluded, "This study presents data suggesting that the changes caused by ozone pollution are long-lasting, and maybe even permanent."[102]

ALA neglected to mention that the monkeys were exposed to 0.5 ppm ozone for eight hours a day for five days in a row, followed by nine days of clean air. This cycle was repeated eight times. To give you an idea of the magnitude of these ozone exposures, during the last thirty years, *only one site in the United States has ever exceeded 0.5 ppm ozone for even an hour, and that happened in 1976.* Today, the worst site in the United States never reaches even 0.25 ppm for an hour, and the average site never reaches 0.11 ppm. In other words, the study ALA cites to imply that ozone is causing permanent lung damage and asthma in children is based on a laboratory study that used a level and frequency of ozone exposure far higher than anyone anywhere in the United States has ever experienced.

ALA can't take all the blame for misstating the results of the UC-Davis study. They had help from the scientists who performed the study. A UC-Davis press release began, "Primate Research Shows Link between Ozone Pollution, Asthma," and claimed the ozone exposures "mimick[ed] the effect of exposure to occasional ozone smog—for example as it occurs in the Sacramento area."[103]

In addition to asthma, the Children's Health Study assessed the relationship between air pollution and growth in children's lung function.[104] After following more than 1,700 children from ages ten to eighteen between 1993 and 2001, the study reported that there was no association between ozone and lung-function growth—this despite the fact that the twelve communities in the study ranged from zero to more than one hundred twenty 8-hour exceedance days per year, and zero to more than seventy 1-hour exceedance days per year during the study period.[105]

No area outside California has ever had anywhere near this frequency of elevated ozone, even for a single year, much less for several years running. If seventy or one hundred twenty ozone exceedance days per year don't reduce kids' lung capacity, then the far lower levels that most Americans experience certainly won't be having an effect either. This hasn't stopped environmental groups from claiming otherwise. For example, in

"Impacts of Ozone on Our Health," the Carolinas Clean Air Coalition claims, "Children have a 10 percent decrease in lung function growth when they grow up in more polluted air."[106]

The Children's Health Study also suggests that $PM_{2.5}$ is causing little long-term harm. Unlike ozone, $PM_{2.5}$ actually was associated with a small effect on lung development. Annual-average levels ranged from about 6 to 32 $\mu g/m^3$ in the twelve communities in the study.[107] Across this range, $PM_{2.5}$ was associated with about a 2.0 percent decrease in forced expiratory volume in one second (FEV_1) and a 1.3 percent decrease in force vital capacity (FVC). Both tests are measures of lung capacity.

But even this drastically inflates the apparent importance of the results, because no location outside of the CHS communities has $PM_{2.5}$ levels anywhere near 32 $\mu g/m^3$. In fact, even the worst area in the United States—Riverside, California, which is one of the CHS communities—averaged 25 $\mu g/m^3$ for 2002–4. There also didn't appear to be any decrease in lung function until average $PM_{2.5}$ levels exceeded about 15 $\mu g/m^3$.[108] But 90 percent of the nation's monitoring locations are already below 15 $\mu g/m^3$.

It is also worth noting that the children in the CHS were already ten years old when they entered the study, and had therefore been breathing the even higher pollutant levels extant during the 1980s in Southern California. For example, Riverside averaged about 48 $\mu g/m^3$ $PM_{2.5}$ during the 1980s, or 50 percent greater than the average from 1994 to 2000, the years during which $PM_{2.5}$ levels were measured for the Children's Health Study analyses.[109] If it was really these higher levels that caused the lung-function declines, then even today's very highest $PM_{2.5}$ levels would be causing no more than a 0.5 percent decrease in children's lung capacity.

Thus, taking the CHS results at face value, ozone is having no effect on children's lung development anywhere in the United States, and $PM_{2.5}$ is having no effect in the vast majority of the country. Even in areas with the highest levels, $PM_{2.5}$ is having virtually no effect on lung development.

Despite these findings, the USC press release on the study created the appearance of serious harm. Titled "Smog May Cause Lifelong Lung

Deficits," it asserted, "By age 18, the lungs of many children who grow up in smoggy areas are underdeveloped and will likely never recover."[110] The National Institutes of Health also made inaccurate claims about the study. In the NIH press release, the director of the National Institute of Environmental Health Sciences claimed the study showed that "current levels of air pollution have adverse effects on lung development in children."[111]

The study itself, which was published in the prestigious *New England Journal of Medicine* (*NEJM*), didn't even explicitly reveal the average percentage change in children's lung function in going from the highest to the lowest $PM_{2.5}$ areas. Instead, readers had to be vigilant enough to realize it could be derived by combining information found in three different places in the article.[112]

The researchers reported a different outcome measure in their *NEJM* paper: the percentage of children in each community with a lung capacity of less than 80 percent of the "predicted" value for their age.[113] $PM_{2.5}$ was associated with a nearly fivefold increase in this percentage, from 1.6 percent of children in the lowest-$PM_{2.5}$ community to 7.9 percent in the highest-$PM_{2.5}$ community.

This seems like a large effect, but it isn't. What's going on is that the 2 percent average decline in lung function in the highest-$PM_{2.5}$ community relative to the lowest meant a shift of some children who were at, say, 80 or 81 percent of "predicted" lung capacity for their age, down to maybe 78 or 79 percent. Because lung-capacity scores have a bell-curve distribution, and few children have low lung capacity, there are many more children slightly above 80 percent than slightly below 80 percent. A small shift in average lung-capacity scores therefore results in a large change in the fraction of children scoring below a given cutoff level.[114]

NIH took advantage of this statistical subtlety in its own press release, which began: "Children who live in polluted communities are five times more likely to have clinically low lung function—less than 80 percent of the lung function expected for their age."[115] This statement creates the appearance of a decline of more than 20 percent in average lung function by leading readers tacitly to assume that all kids would be at 100 percent if there were no air pollution.

Environmentalists have, indeed, mistakenly taken this statement to mean that children lost 20 percent of their lung function. For example, in discussing the results of the study, the American Lung Association's *State of the Air: 2005* report claims the "average drop in lung function was 20 percent below what was expected for the child's age."[116] And the Carolina's Clean Air Coalition (CCAC) said, "Medical studies show that children who grow up in areas as polluted as the Charlotte [North Carolina] region are losing up to 20 percent of their lung function—permanently."[117]

We asked the Carolinas Clean Air Coalition for their source for this claim. The CCAC's director sent us a copy of the NIH press release discussed above. That's not the only thing the CCAC got wrong. Recall that the decline in lung function was based on $PM_{2.5}$ levels averaging more than 30 $\mu g/m^3$. But the worst areas of the Charlotte region average about 15 or 16 $\mu g/m^3$, or half the peak levels of the Children's Health Study.

Many portrayals of the long-term health risks from air pollution leave out crucial contextual information on the size of the effects and the pollution levels necessary to cause the effects. If you ignore this crucial information, then you can say without qualification, as a Sacramento air quality regulator did, that "ground-level ozone destroys certain parts of your lungs."[118] That's a true statement, but only for ozone levels several times higher and exposures many times longer than ever occur in Sacramento or anywhere else in the United States.

Does Air Pollution Aggravate Preexisting Health Problems?

Air pollution can aggravate preexisting respiratory diseases, resulting in anything from mild discomfort to a trip to the emergency room. As for other health effects, the data do not support claims of serious health problems.

For example, when EPA developed the 8-hour ozone standard in 1996, the agency estimated that going from full national attainment of the 1-hour standard to attainment of the 8-hour standard would reduce hospitalizations for asthma attacks by 0.6 percent.[119] EPA buried this finding of small benefits in the middle of a technical report in the *Federal Register* and never mentioned it anywhere else.

More recent official estimates have only borne out the conclusion that the overall effects of air pollution are small. In a 2005 study published in the journal *Environmental Health Perspectives*, EPA scientists estimated that reducing nationwide ozone from 2002 levels, which were by far the highest since 1999, to the federal 8-hour standard would reduce asthma emergency room visits by 0.04 percent, and respiratory hospital admissions by 0.07 percent.[120] In other words, based on EPA's estimates, going from the relatively high ozone levels of 2002 down to the 8-hour standard would prevent 1 of every 2,500 asthma ER visits and 1 of every 1,400 respiratory hospital admissions.

The California Air Resources Board recently adopted an ozone standard for California that is much tougher than the federal standard, requiring ozone to be reduced near or even below background levels across the state.[121] Despite the fact that millions of Californians live in areas with much higher ozone than the rest of the country, CARB predicts that reducing ozone will result in little health improvement. For example, based on CARB's estimates, going from 2001–3 levels down to 0.070 ppm—the level of CARB's standard, and, in effect, an elimination of all human-caused ozone in the state—would reduce emergency room visits for asthma by 0.35 percent, and respiratory-related hospital admissions by 0.23 percent.[122] In other words, eliminating all human-caused ozone would prevent about 1 of every 450 asthma ER visits and 1 of every 300 respiratory hospital admissions. If health benefits also accrue when ozone is reduced from levels that already comply with the 0.070 ppm standard, CARB predicts the benefits of attaining the standard will be about five times greater.[123]

How can CARB, EPA, and environmentalists estimate that air pollution has a tiny quantitative effect on public health, but then create an impression of widespread harm in their rhetoric? They simply don't publicize the quantitative estimates, and sometimes don't even calculate them explicitly. The CARB and EPA studies cited above don't contain any explicit estimates of the percentage reduction in adverse health effects expected from ozone reductions. Instead, they provide only the number of cases of a given health effect predicted to be prevented by reducing ozone.

For example, EPA estimated that going from 2002 ozone levels to full national attainment of the 8-hour standard would prevent 6,500 respiratory hospital admissions each year.[124] CARB estimated that statewide attainment of its much tougher standard would prevent more than 800 respiratory hospital admissions each year in California.[125]

To convert these numbers into percentages, you need to divide them by the *total* number of respiratory hospital admissions each year. Fortunately, both EPA and CARB provide estimates of the rate of respiratory hospital admissions—that is, the number of respiratory hospital admissions per capita.[126] You can multiply this *rate* by the total population to get the total number of respiratory hospital admissions. EPA and CARB don't do this calculation themselves. One might surmise that this is because it would be embarrassing for these agencies to admit publicly that even they themselves have concluded that air pollution is responsible for at most 1 or 2 percent of all respiratory and cardiovascular distress.

Even alarming reports sometimes include buried information that moderates or even refutes the scary headlines. For example, the Clean Air Task Force (CATF) has produced a number of reports with titles like *Power to Kill* and *Death, Disease and Dirty Power*.[127] These reports draw their main claims from a more sober study funded by CATF and produced by consultants from Abt Associates.[128] *The Particulate-Related Health Benefits of Reducing Power Plant Emissions* concludes that a complete elimination of all power plant emissions in the United States would reduce serious respiratory distress, such as emergency room visits for asthma, pneumonia, or cardiovascular disease, by 0.4 to 1.6 percent.[129] Yet power plants contribute about one-third of all $PM_{2.5}$ in the eastern half of the U.S.[130] These results are reported in a table near the back of the Abt report and don't appear in any of the derivative and more sensational reports that CATF produced for public and media consumption. In other words, environmentalists create the impression that power plants are responsible for a large fraction of all respiratory disease and distress. But even the technical studies they commission conclude that power plants impose a relatively small health burden.

In any case, the harm environmentalists claim for power plant pollution isn't real. The main form of particulate matter from coal-fired power plants is ammonium sulfate, formed from sulfur dioxide emissions, as well as smaller amounts of ammonium nitrate formed from NOx.[131] But laboratory studies with human volunteers, including volunteers with respiratory diseases, have shown that sulfate and nitrate are not toxic, even at levels many times the maximum levels found in ambient air.[132] In fact, ammonium sulfate has been used as an inert control—that is, a harmless compound—in studies of the health effects of acidic aerosols.[133] Inhaled magnesium sulfate is used therapeutically to *reduce* airway constriction in asthmatics.[134] Because sulfates and nitrates are not toxic, environmentalists are mistaken when they claim that reducing particulate matter from power plants would have any health benefits.

The small benefits regulators estimate from ozone reductions are also likely to be overestimates, because they are based on a selective reading of the health effects literature that ignores contrary evidence.

For example, researchers from Kaiser Permanente studied the relationship between air pollution and emergency room visits and hospitalizations in California's Central Valley and reported that higher ozone was associated with a statistically significant *decrease* in these health effects.[135] CARB sponsored this study, but did not cite or include it when estimating the ostensible benefits of a tougher ozone standard.[136] EPA also failed to mention this study in its latest review of the federal standard.[137] This selective use of evidence creates the impression that air pollution's effects are larger and more certain than suggested by the overall results of health effects research.[138]

Providing the percentage reduction in a given health effect achievable through pollution reduction gives the public a better idea of the total burden of disease that can be avoided by reducing air pollution and puts the harm from pollution into the appropriate context of the overall burden of disease. Doing this shows that even if we take regulators' and environmentalists' estimates at face value, air pollution is having a tiny effect on Americans' health. Taking account of the contrary evidence omitted by regulators and environmentalists reduces the purported harm from air pollution still further.

A range of other studies and health data show that air pollution has at worst a minor effect on Americans' health. For example, a recent study of children in Connecticut and western Massachusetts followed 271 children with asthma from April through September of 2001 and compared various symptom rates with ozone and $PM_{2.5}$ levels. It concluded that higher ozone was associated with increased asthma symptoms, such as shortness of breath and wheeze, as well as bronchodilator (inhaler) use, even at levels substantially below the current federal 1-hour and 8-hour standards.[139] These effects were found for asthmatic children who were on maintenance medication—that is, more severe asthmatics—but not those with milder asthma. $PM_{2.5}$ had no association with asthma symptoms.

The National Institutes of Health, which funded the study, put out a press release highlighting the findings that an increase of 0.05 ppm in 1-hour ozone levels was associated with a 35 percent increase in the prevalence of wheezing and a 47 percent increase in the prevalence of chest tightness in asthma sufferers who used maintenance medication, but not in those who were not on medication.[140] The lead researcher was quoted in the press release as saying, "Our results suggest that ground-level ozone is strongly associated with adverse health effects in children with asthma, even at levels below the current federal standards."

These percentages suggest big effects, and they are supported by a definitive statement from a government-sponsored medical authority. But neither the study itself nor the press release assessed the real-world importance of the findings or their policy implications. In reality, the study shows that even if the results are taken at face value, large reductions in ozone would result in small health improvements for asthmatics.

Medication users, those for whom a pollution association was reported, experienced wheezing on 2.8 percent of days and chest tightness on 1.2 percent of days.[141] Thus, on a day when 1-hour ozone jumped 0.05 ppm, we would expect the risk of wheezing to rise 35 percent, from 2.8 to 3.8 percent, and the risk of chest tightness to rise 47 percent, from 1.2 to 1.8 percent. So ozone was associated with a substantial percentage rise in symptoms. But the symptoms were relatively uncommon to begin with. For example, a 0.05 ppm rise in 1-hour ozone was associated with an increase in the fraction of children experiencing wheezing from one in thirty-six to one in twenty-six.

We can also ask how much we should expect asthma symptoms to decline due to reductions in ozone, once again taking the study results at face value. Assuming the background level for peak daily 1-hour ozone is 0.04 ppm, average daily 1-hour levels were 0.027 ppm above background during the study period. If these increases were completely eliminated, the risk of experiencing wheezing on a typical spring or summer day would drop from 2.4 to 1.9 percent. The risk of chest tightness would drop from

1.2 to 0.9 percent. These are significant percentage changes in risk, but small absolute changes. And attaining these benefits would require reducing ozone to background levels. Merely attaining the 8-hour standard would have much smaller benefits, reducing the risk of wheezing from 2.4 to 2.3 percent and chest tightness from 1.2 to 1.1 percent.

The analysis above assumes that the study uncovered a genuine causal connection between ozone and asthma symptoms. But this was an observational study, not a randomized trial, and the study also had a number of serious technical flaws. For example, instead of using the nearest pollution monitor to represent exposure for a given child, the researchers averaged pollution levels at all monitors, which were spread over a few thousand square miles of Connecticut and western Massachusetts, where the study was performed. So the study's ozone exposure measure had little to do with the actual levels experienced by any particular child in the study.

Moreover, the statistical model used inappropriate controls for confounding by weather. For example, the model included same-day weather but not weather from previous days, even though previous days' weather could have a delayed effect on health. This is particularly important, because the researchers did include air pollution effects from previous days in their statistical model. Since air pollution and weather can be correlated, effects the researchers attributed to the one could have been due to the other. Another problem is that the only weather variable included in the statistical model was temperature, even though other aspects of meteorology can affect asthma symptoms. The study also did not control for day of the week and season, which also could confound the results since, for example, asthma symptoms rise in September, independent of pollution levels.[142]

The two studies just discussed, the Kaiser Permanente study in California's Central Valley and the New England study, also show how the overall evidence in the research literature is far more equivocal than popular portrayals suggest. The Central Valley study reported harm from particulate matter, but not ozone. The Connecticut study reported harm from ozone, but not particulate matter. Regulators, health scientists, and environmentalists generally highlight only the PM results from the Central Valley study and only the ozone results from the New England study, creating an appearance of consistency and robustness in the research base that does not in fact exist.

This chapter has already presented several lines of evidence that air pollution does not cause people to develop asthma and is at worst a minor factor in the exacerbation of preexisting asthma. Another source of evidence for the unimportance of air pollution is the pattern of hospital visits for asthma. Emergency room visits and hospitalizations for asthma are *lowest during July and August, when ozone levels are at their highest*.[143] If ozone were a significant cause of asthma exacerbations, this is just the opposite of what we would expect to find.

State respiratory hospitalization data also conflict with the claim that ozone exacerbates asthma. For example, North Carolina counties with the *lowest* ozone levels have the *highest* rate of asthma hospitalizations. This is shown in figure 7-3. Each graph represents an individual year, and each point represents a North Carolina county. The vertical axis gives the number of 8-hour ozone exceedance days in that year. For counties with more than one monitoring site, the ozone value is an average of all sites in the county. The horizontal axis gives the number of asthma hospitalizations per 100,000 population. The lines through the data points are linear regression lines. Note that counties with the *lowest* ozone have the *highest* hospitalization rates.

UCLA's California Health Information Survey (CHIS) reported the percentage of children in each California county who had been previously diagnosed with asthma and who reported experiencing symptoms in 2001.[144] Figures 7-4 and 7-5 on pages 156 and 157 compare the CHIS asthma data with ozone and PM_{10} levels in California counties for the same year. Note that there is little relationship between the two. Some counties with high pollution levels have low asthma rates, while some with low pollution levels have high asthma rates.

Even asthma experts seem to be unaware of key facts about the illness. For example, some doctors claim they can tell that ozone causes asthma attacks by the number of asthma sufferers coming into their clinics. Here's one example from an NIH press release: "[Jesse] Joad [director of the UC Davis Pediatric Asthma program] noticed that when air quality goes down, admissions with severe asthma go up. 'We had a really bad week of admissions in mid-September,' she says, 'when there were high ozone days at the beginning of the week.'"[145] But, as already noted, asthma attacks rise in September all over the world, regardless of pollution levels.[146] If high

FIGURE 7-3

ASTHMA HOSPITALIZATION RATE VS. OZONE LEVELS IN NORTH CAROLINA COUNTIES

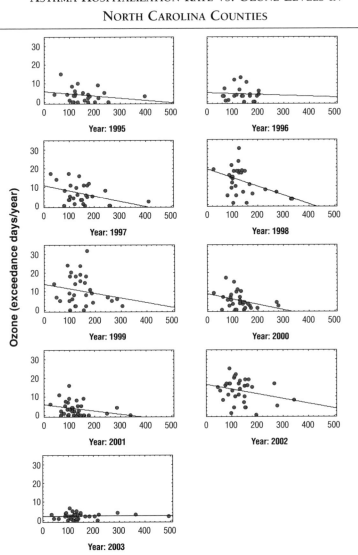

SOURCES: Ozone data for North Carolina were downloaded from EPA's AIRData database, http://www.epa.gov/air/data/geosel.html (accessed September 29, 2006). Asthma hospitalization data were provided by the North Carolina State Center for Health Statistics.

NOTES: Each point represents a given North Carolina county. Ozone exceedance days are based on the 8-hour standard and are an average for all monitoring sites operating in a given county in a given year.

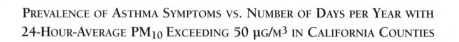

FIGURE 7-4

PREVALENCE OF ASTHMA SYMPTOMS VS. NUMBER OF DAYS PER YEAR WITH
24-HOUR-AVERAGE PM$_{10}$ EXCEEDING 50 μG/M^3 IN CALIFORNIA COUNTIES

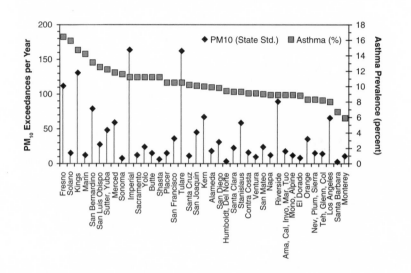

SOURCES: CARB, *2006 Air Pollution Data CD*, February 2006, http://www.arb.ca.gov/aqd/aqdcd/aqdcd.htm
(accessed April 2, 2006); UCLA Center for Health Policy Research, "Asthma Symptom Prevalence in
California in 2001," in *2001 Health Interview Survey*, Spring/Summer 2002, http://www.healthpolicy.
ucla.edu/pubs/files/Asthma-by-county-052002.pdf (accessed September 18, 2006).

NOTES: Number of days exceeding the pollution standard is an unweighted average for all monitoring sites
in a county with data for 2001. Asthma prevalence by county for 2001 is the percentage of children age
zero to seventeen years who reported both having been previously diagnosed with asthma and experiencing asthma symptoms in 2001. California's state 24-hour PM$_{10}$ standard is 50 μg/m^3. It is substantially
more stringent than the federal 24-hour PM$_{10}$ standard of 150 μg/m^3.

ozone could cause a noticeable increase in asthma exacerbations, it would
be obvious in July and August, when ozone levels are at their highest in
the Northern Hemisphere. Yet all over the United States, asthma-related
emergency room visits are at their *lowest* in July and August.

Asthma can be a serious and frightening disease, and nobody wants to
see children or anyone suffering from it. The point isn't whether it would
be good to reduce the harm from asthma. Of course it would. But policy-

Figure 7-5

Prevalence of Asthma Symptoms vs. Number of Days per Year with Peak 1-Hour Ozone Exceeding 0.095 ppm in California Counties

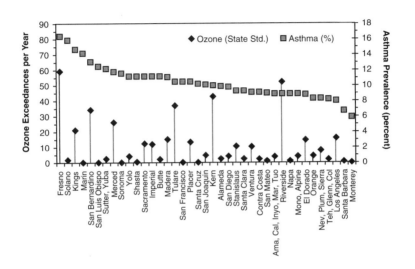

Sources: CARB, *2006 Air Pollution Data CD*, February 2006, http://www.arb.ca.gov/aqd/aqdcd/aqdcd.htm (accessed April 2, 2006); UCLA Center for Health Policy Research, "Asthma Symptom Prevalence in California in 2001," in *2001 Health Interview Survey*, Spring/Summer 2002, http://www.healthpolicy.ucla.edu/pubs/files/Asthma-by-county-052002.pdf (accessed September 18, 2006).

Notes: Number of days exceeding the pollution standard is an unweighted average for all monitoring sites in a county with data for 2001. Asthma prevalence by county for 2001 is the percentage of children ages zero to seventeen years who reported both having been previously diagnosed with asthma and experiencing asthma symptoms in 2001. California's state 1-hour ozone standard is 0.095 ppm. It is slightly more stringent than the federal 8-hour standard.

makers must ask whether seeking large additional reductions in ozone levels that are already relatively low is the best or even a good way to help asthmatics. It would cost hundreds of billions of dollars per year on a nationwide basis to reduce ozone from current levels to background levels (which would require going well below the already stringent 8-hour standard and an elimination of all human-caused ozone-forming emissions), assuming that goal is even achievable. At best, these reductions would result in tiny and imperceptible benefits for asthma sufferers and perhaps no health benefits at all.

A number of health researchers and regulators have claimed that higher air pollution levels increase school absences. But here again, the evidence suggests the two are unrelated. For example, the California Air Resources Board predicts that going from ozone levels during 2001–3 down to the state's new 8-hour standard of 0.070 ppm—about a 40–50 percent reduction in peak ozone levels in most of the state—would reduce total school absences by nearly 9 percent.[147] That amounts to a reduction of 3.7 million absences per year, or a little over half an absence day per student per year.[148]

Whatever the benefits of ozone reductions in terms of school absences, they would presumably apply to absences related to respiratory illnesses, rather than to, say, gastrointestinal illnesses or non–illness-related absences. Data collected in the Children's Health Study suggest that respiratory illness is a factor in about 35 percent of school absences.[149] If so, then CARB implicitly predicts that attaining its 8-hour standard will reduce respiratory-related absences by 26 percent.[150]

This prediction is not credible. First, even without examining the analytical merits of the estimate itself, comparison with CARB's other health-effects estimates raises concerns about consistency. CARB predicts that reducing ozone is seventeen and twenty-two times as effective in reducing respiratory-related school absences as in reducing, respectively, "minor restricted-activity days" and respiratory hospital admissions.[151] The degree of sickness necessary to cause an absence from school would presumably fall somewhere between the extremes of the other two illnesses, making CARB's estimate for ozone and school absences inconsistent with those for the other illnesses.

Second, CARB ignores the biological implausibility of the results in the study on which it based its estimates of ozone's effect on school absences.[152] F. D. Gilliland and colleagues reported that a 0.020 ppm increase in 8-hour ozone levels was associated with an 83 percent increase in school absences due to respiratory causes. However, the apparent effects were due mainly to ozone levels from one or two weeks previously, rather than during the past few days. Furthermore, spending *more* time outdoors, which would have increased ozone exposures, was associated with *fewer* school absences.

When Gilliland and colleagues assessed PM_{10} and school absences, they concluded that PM_{10} was associated with a large increase in *non*–illness-related absences, but not with absences due to illness. An increase of ten micrograms per cubic meter in PM_{10} was associated with a larger effect on non–illness-related absences than was a 0.020 ppm increase in ozone on respiratory-related absences.

Taken together, these results are biologically implausible and suggest that the apparent effect of ozone on school absences was a statistical figment, possibly due to the researchers' failure to control adequately for season, rather than a real cause and effect relationship.

Third, other studies contradict CARB's analysis. Gilliland et al. (2001)—the study CARB relied on—used Children's Health Study data to assess the relationship between school absences and ozone. But there are two other studies—one by K. Berhane and D. C. Thomas from 2002 and another by V. Rondeau and colleagues from 2005—that used the same CHS data and included some of the same researchers as co-authors, but reported no statistically significant association between daily ozone levels and school absences.[153]

Berhane and Thomas (2002) and Rondeau et al. (2005) provide additional information that casts doubt on the claim of an air pollution–school absence linkage. For example, both show that the apparent effect of short-term ozone levels on respiratory absences increases as the "lag time"—that is, the number of days between a given ozone exposure and an absence from school—included in the statistical model increases. When looking at only the last five days of ozone exposures before an absence, Rondeau and colleagues reported that *higher* ozone was actually associated with a small, statistically insignificant *decrease* in school absences. Ozone was associated with an insignificant increase in absences once the ozone-exposure lag-time was increased to fifteen days, and the effect increased a bit more at a thirty-day lag, though it was still small and well below statistical significance.

Berhane and Thomas reported no ozone association with illness-related absences over a fifteen-day lag-period, but they did report a nearly significant association for a thirty-day lag-period. As noted earlier, it seems implausible that ozone exposures from weeks prior would have a greater effect on school absences than exposures during the previous few days.

While Gilliland et al. and Rondeau et al. present only aggregate results for the twelve CHS communities, Berhane and Thomas also present results for each of the communities individually, revealing great intercommunity variation. For example, ozone exposures from up to a few days before were associated with a large increase in absences in a few communities, a large decrease in others, and little change in still others. Ozone exposures from one or two or three weeks ago had a similar range of apparent effects. These wild variations among communities in health effects versus time-since-ozone-exposure are not biologically plausible, and once again suggest that the studies are turning up chance statistical correlations rather than real ozone effects.

These three school-absence studies once again illustrate the pitfalls of observational studies. They all used the same data but got wildly different results. And all the results are, in one way or another, implausible.

Laboratory Studies and Short-Term Air Pollution Health Effects

While laboratory studies with human volunteers can't, of course, directly assess the rate of serious health effects from air pollution, they can help test the biological plausibility of the results from observational epidemiology studies. But laboratory studies provide little support for the claims of serious harm from air pollution. We've already discussed the lack of support for harm from sulfate or nitrate particulate matter. Laboratory studies also provide little evidence of acute effects from motor-vehicle-related PM or PM in general—at least at levels representative of real-world exposures.

For example, a Health Effects Institute study exposed healthy and asthmatic volunteers to 200 µg/m^3 of concentrated ambient PM$_{2.5}$ collected in the Los Angeles area. The exposures lasted for two hours, and the subjects exercised intermittently to increase their respiration rate and therefore their PM$_{2.5}$ exposures.[154] This represents a "worst-case" real-world exposure. Even areas with the highest levels in the country average less than 25 µg/m^3 of PM$_{2.5}$, and peak hourly levels rarely exceed even 100 µg/m^3.[155]

Despite the relatively high PM$_{2.5}$ levels in the study, there were no changes in symptoms or lung function in either the healthy or asthmatic subjects, and little evidence of inflammatory responses. Since many inflammatory markers were measured and only a few changed, the authors pointed out that these changes could be due to chance.

Another HEI study exposed both healthy and asthmatic volunteers to 100 µg/m^3 of diesel soot for two hours while they exercised intermittently on a stationary bicycle.[156] The researchers found little evidence of an inflammatory response, and the healthy subjects exhibited more evidence of inflammation than the asthmatics. In fact, according to the project summary, the study "did not find inflammatory changes in asthmatic participants after controlled exposure to diesel exhaust."[157]

Dozens of laboratory studies have assessed people's responses to ozone exposure, and regulators use these as a main justification for further tightening ozone standards. For example, when CARB recently adopted a stringent new 8-hour standard for California (0.070 ppm 8-hour average, no exceedances allowed), the agency's justification rested largely on the claim that laboratory studies demonstrated adverse effects of ozone at levels below the current federal standard: "Our recommendation for the 8-hour standard is based primarily on the chamber [that is, laboratory] studies that have been conducted over the last 15 years, supported by the important health outcomes reported in many of the epidemiologic studies."[158]

And indeed, laboratory studies with human volunteers have provided a great deal of evidence that multihour exposure to ozone while exercising can result in temporary reductions in lung function. For exposures of one to two hours with exercise, these studies report small average reductions in lung function at ozone levels around 0.12 ppm or above—in other words, levels at or above the 1-hour standard.

Ozone levels as low as 0.08 ppm—below the federal 8-hour standard—are associated with lung-function reductions when people are exposed for more than five hours while exercising nearly continuously.[159] Although the average reduction in lung function was small in these studies—average lung capacity declined up to several percent—a few people experienced larger lung-capacity reductions. Effects of these short-term exposures were temporary, and lung function returned to normal within a day.

The problem with these studies is that despite superficial appearances, they are based on unrealistically high ozone exposures. It turns out that ozone concentrations measured at the ambient monitors used to determine Clean Air Act compliance are much higher—at least 65 percent higher, on average—than the concentrations in the air people actually breathe in.[160] Several factors contribute to the discrepancy between monitored ozone levels and personal exposures. Ambient monitors are often placed several feet above typical human head-height to avoid interferences from people and surfaces near the ground. However, ozone deposition on surfaces (such as clothing or the ground) reduces the levels in the air that people actually breathe in. Levels also tend to be lower near roads, due to destruction by nitric oxide emitted by vehicles. Finally, there is evidence that the equipment used for regulatory monitoring gives ozone readings that might be biased high.[161]

However, the ozone doses used in the laboratory studies are based on *ambient concentrations measured by monitors*, rather than real personal exposures. So, for example, a laboratory study that has volunteers breathe air containing 0.08 ppm of ozone is really simulating an ambient ozone level (as measured by Clean Air Act compliance monitors) of around 0.13 ppm (65 percent higher than the personal exposure level) rather than 0.08 ppm.

In addition to using personal exposures that are too high, laboratory studies also use "background" ozone exposures that are too low. To determine the health effects of ozone, researchers compare subjects' lung function while breathing ozone with their lung function while breathing "clean" air—that is, air representing some background exposure level. All studies to date have used ozone-free air for this background level. This too is unrealistic, because there is always some natural background ozone in air due to natural emissions of ozone-forming pollutants from vegetation, lightning, and occasional transport of ozone to ground level from the stratosphere. Some ozone and ozone-forming pollutants are also transported into the United States from other countries.

This background level of ozone is a matter of controversy, but it is certainly not zero. CARB assumed a background of 0.04 ppm when developing its recently adopted 8-hour ozone standard for California.[162] The actual background level may indeed be even greater.[163] In any case, if we assume 0.04 ppm is a realistic background level, then rather than using ozone-free air, laboratory studies should be using a personal-exposure concentration of about 0.025 ppm (to account for the difference between ambient levels and personal exposures) as the background exposure when assessing the health effects of elevated ozone.

To summarize, a typical laboratory study compares people's lung function when breathing ozone-free air with, say, 0.08 ppm ozone. Researchers and regulators consider such a study to be comparing ozone at a level near the 8-hour standard of 0.085 ppm with unpolluted air. But in fact the study is doing nothing of the sort. In terms of what would be measured at Clean

Air Act compliance monitors, it is really comparing ozone at a level greater than even the 1-hour standard (0.125 ppm) with an ozone level of zero.

To assess realistically the effects of an ambient ozone level of 0.08 ppm (as measured at Clean Air Act compliance monitors), researchers would need to expose people to 0.025 ppm ozone to represent "clean air" and then expose them to 0.05 ppm to represent an ambient level of 0.08 ppm. No one has yet done a laboratory study of the effects of such a small change in ozone levels. However, one study compared levels of 0.00, 0.04, and 0.08 ppm.[164] College students were exposed to these levels during vigorous exercise (on a treadmill for fifty minutes of every hour) for 6.6 hours. The vigorous exercise increased exposure to several times the level that would occur for people at rest or walking. There was no statistically significant difference in average lung-function tests between the 0.00 and 0.04 ppm exposures. Average test scores declined about 5 percent between the 0.04 ppm to 0.08 ppm exposures, though this difference appeared only after six hours of exposure.

Recall that personal exposures of 0.04 ppm and 0.08 ppm are equivalent to ambient-monitor levels of 0.06 ppm and 0.13 ppm. Yet across this range of levels, the latter of which is higher than even the 1-hour standard, it took more than five hours of vigorous exercise to elicit even a 5 percent change in lung function. Based on these results it seems unlikely that people would experience any change in lung function in going from a background personal ozone exposure of 0.025 ppm to the 0.05 ppm level representative of personal exposures when ambient ozone reaches the level of the current federal 8-hour standard.

These laboratory studies are unrealistic in additional ways. For example, regulators intend their air pollution standards to protect the most "sensitive" groups, which they consider to be children, the elderly, and people with respiratory diseases. But the reason these laboratory studies are done with college students is that people in the "sensitive" groups can't maintain the vigorous physical activity levels for the several hours necessary to elicit measurable health effects at ozone levels that might be encountered in the real world.

Thus, rather than supporting tougher standards, the laboratory studies suggest that even the current federal 8-hour standard of 0.085 ppm is more than stringent enough to protect people, even sensitive people, from harmful effects due to ozone exposure.

Does Air Pollution Cause Cancer?

Even in high-pollution areas, regulators estimate the cancer risks from pollution to be relatively small. For example, MATES-II, a major study by regulators in Southern California, estimated the lifetime risk of developing cancer from air pollution at about 1 in 600, or 0.17 percent, in areas with the highest pollution levels.[165]

Given this estimate, what fraction of all cancers could be caused by air pollution? The average person's lifetime risk of developing cancer is about 42 percent—that is, about four of every ten people will develop cancer at some point in their lives.[166] If so, then the fraction of all cancers that are due to air pollution is 0.17/42, or about 0.4 percent. Put another way, assuming the air pollution cancer risk estimate is correct, if you've developed cancer, there's a 1 in 250 chance it was caused by air pollution. This is consistent with other estimates of the fraction of all cancers due to exposure to manmade chemicals.[167]

Still, even at this risk level, air pollution would be causing several thousand new cancer cases per year.[168] If true, this would be a tragedy. But this estimate was based on an assumption that everyone in the United States is exposed to the same level of carcinogenic air pollutants as people in the worst areas of Los Angeles when, in fact, most are exposed to much lower levels.

Furthermore, the risk estimates themselves are implausible, because they are based on false assumptions that drastically inflate the risk. The first false assumption is that the risks of cancer due to very high chemical exposures can be extrapolated down to comparatively minuscule real-world ambient pollution levels. The second false assumption is that people are exposed to the highest measured ambient levels twenty-four hours a day, seven days a week, for seventy years.

Low doses of chemicals are unlikely to cause cancer.[169] Animal studies of the ability of chemicals to cause cancer are done using the so-called maximum tolerated dose (MTD). The MTD is the highest dose of a chemical that you can give to an animal without killing it. Just over half of all chemicals that have ever been tested, *both natural and synthetic*, are carcinogenic in rodents based on the MTD, including about half the chemicals naturally present in coffee that have been tested so far.[170]

Regulators' cancer risk estimates are based on extrapolation of these high-dose studies down to the comparatively minuscule exposures in the real world, on the assumption that the cancer risk per unit of chemical is the same at low doses as at high doses. But very high doses of chemicals cause chronic tissue inflammation, which is itself a risk factor for cancer and may explain why many chemicals cause cancer at very high doses but are unlikely to be carcinogenic at low doses.[171] Another factor that argues against low doses causing cancer is that people have defense mechanisms against carcinogens, and the activity of these mechanisms increases with increasing exposure to chemicals.[172]

The second assumption is definitely false. First, few people spend all of their time in one place, and they spend much of their time indoors, where pollution from outdoor sources may be lower. Second, as we saw earlier, ambient levels of all kinds of pollutants, including high-dose carcinogens, have been dropping and will continue to drop. So no one could possibly be exposed to recent pollution levels for seventy more years, or even ten more.

In any case, the MATES-II study estimated that diesel soot accounts for about 70 percent of the cancer risk from air toxics, with benzene and 1,3-butadiene contributing another 10 percent each. All other air pollutants combined accounted for the remaining 10 percent. Thus, 90 percent of the purported cancer risk from air pollution is due to just three pollutants. Yet, as shown earlier, requirements that have already been adopted, or, in some cases, implemented, will reduce automobile VOC emissions, including benzene and 1,3-butadiene, and diesel soot emissions at least 80 percent during the next twenty years or so, even after accounting for growth in total miles driven by motor vehicles. Thus, whatever the cancer risks from current air pollution emissions, they will be largely eliminated by already-adopted requirements.

Health Benefits from Air Pollution?

Ozone up in the stratosphere protects us from the sun's ultraviolet (UV) rays, which can cause skin cancer and cataracts at high enough exposures. Ozone near ground level, including human-caused ozone, adds a small additional increment to this protection. Even without human-caused ozone, about 10 percent of the total ozone above our heads is found within ten miles of sea level.[173] Reducing ozone in the lower atmosphere, including ozone caused by human activities, thus has the potential to increase people's exposure to UV light and thereby increase their risk of developing skin cancer and cataracts.

An internal EPA analysis in 1997 concluded that the incremental ozone reductions (beyond the federal 1-hour ozone standard) necessary to attain the 8-hour standard would result in an additional seven hundred cases of nonmelanoma skin cancer each year nationwide, due to increased exposure to solar UV light. EPA never officially made this analysis public.[174] Estimates by the Department of Energy (DOE) suggest that reducing ozone to the 8-hour standard would also result in a few thousand additional cases of cataracts, a few dozen cases of melanoma skin cancer, and several melanoma deaths each year.[175]

Despite the eagerness of California and federal environmental regulators to regulate tiny or perhaps nonexistent risks in other circumstances, both EPA and CARB have avoided acknowledging the potential increases in cancer and cataracts due to lower ground-level ozone levels. For example, in a recent extensive analysis of health effects preparatory to further tightening the 8-hour standard, EPA had this to say about the effects of ozone reductions on Americans' UV exposure:

> Within the uncertain context of presently available information on UV-B surface fluxes, a risk assessment of UV-B-related health effects would need to factor in human habits (e.g., daily activities, recreation, dress, and skin care) in order to adequately estimate UV-B exposure levels. Little is known about the impact of variability in these human factors on individual exposure to UV radiation. Furthermore, detailed information does not exist regarding the relevant type (e.g., peak or cumulative) and time period (e.g., childhood, lifetime, or current) of exposure, wavelength dependency of biological responses, and interindividual variability in UV resistance. . . . In conclusion, the effect of changes in surface-level O_3 concentrations on UV-induced health outcomes cannot yet be critically assessed within reasonable uncertainty.[176]

It is odd that EPA would claim *uncertainty* as a reason *not* to address a potential health risk. EPA normally uses uncertainty as the justification for more stringent regulatory limits. And the tenuousness of EPA's risk claims has not prevented the agency from imposing stringent regulations in other contexts.

EPA's uncertainty claim is mistaken in any case. Back in 1997, an internal EPA analysis concluded that "any decrease in atmospheric ozone (tropospheric or stratospheric) causes . . . an increase in the incidence of non-melanoma skin cancers. . . . The methodology for estimating such increases (of both UV levels and skin cancer incidence) is well established."[177]

The reductions necessary to attain the current 8-hour ozone standard could easily result in a few thousand new cases of skin cancer and tens of thousands of cases of cataracts each year.[178] The much larger reductions necessary to attain the more stringent standards EPA is now considering could double or treble this toll. The potential harms from lower ozone levels are sufficient to offset much or perhaps even all of the benefits from further reductions, depending on how large those benefits really are.

The California Air Resources Board also tied itself up in contradictions while trying to avoid addressing the UV-exposure effects of lower ground-level ozone. The agency claimed that the effect of any increase in UV exposure would be "very small" and therefore unworthy of regulatory attention.[179] But CARB has not had similar reservations about regulating other very small risks. For example, like EPA, CARB considers a cancer risk to be unacceptable if the predicted risk is greater than one in one million over a lifetime.[180] And this risk level is determined by methods that by their very nature greatly exaggerate cancer risks (see discussion of this in previous subsection). But the ozone reductions necessary to attain CARB's California 8-hour standard could easily cause an increase in cancer risk of ten in one million, or ten times the agency's own threshold level for regulating other purported cancer risks.[181]

Mercury and Health

Mercury differs from other air pollutants in that the vast majority of exposure to humans and wildlife results not from inhalation of mercury in the air, where it is present at vanishingly small levels, but from eating fish contaminated with an organic form of mercury called methylmercury (MeHg).[182] Mercury in fish ultimately comes from a combination of human-caused and natural mercury emissions to the environment.[183]

There is no question that high levels of mercury can cause serious harm to the brain and nervous system. Tragic poisonings in Japan in the 1950s and '60s and in Iraq in the early '70s demonstrated the danger of high mercury exposures.[184] Highly exposed children suffered mental retardation, cerebral palsy, and seizures. But these episodes involved mercury levels tens to hundreds of times greater than even relatively highly exposed Americans ever experience. Furthermore, where Americans are exposed to tiny amounts of mercury over a long period of time, the poisoning incidents involve very large exposures over a short period of time.

The key question for people in the United States is whether the trace exposures experienced by Americans could be causing harm. In particular, could MeHg in fish consumed by pregnant women be causing later cognitive and neurological damage to children exposed to MeHg in the womb? This question achieved currency a few years ago, when studies by the Centers for Disease Control suggested that about 8 percent of women of childbearing age, defined by the CDC as ages sixteen to forty-nine in this study, had blood mercury levels greater than EPA's "reference dose" (RfD). The RfD is the daily exposure level of a given chemical that NIH estimates is "likely to be without an appreciable risk of deleterious effects during a lifetime."[185]

Based on these data, EPA initially estimated that 320,000 babies are born each year who have been exposed to mercury at levels above the RfD. But EPA revised the estimate to 630,000 in 2004, based on evidence that MeHg was more concentrated in umbilical cord blood when compared with maternal blood.[186] Based on the exposure data and the fact that an estimated 41 percent of U.S. mercury emissions come from the burning of coal for electricity, activists and a number of medical experts have claimed that hundreds of thousands of American children are born brain-damaged or mentally retarded each year due to prenatal mercury exposure, and have called for large reductions in mercury emissions from coal-fired power plants.[187] But the data do not support these claims.

First, actual mercury exposure for pregnant women is far lower than EPA claims. EPA initially claimed that 8 percent of newborns have been exposed to mercury exceeding the RfD and then doubled that figure to

16 percent based on the higher concentration of MeHg in umbilical cord blood relative to maternal blood. These figures were based on women ages sixteen to forty-nine measured in 1999–2000 in the National Health and Nutrition Examination Survey (NHANES) conducted by the CDC.

But NHANES also examined pregnant women specifically, and their average mercury level was 35 percent lower than the average for women ages sixteen to forty-nine.[188] One reason pregnant women's mercury exposures are lower is that blood mercury levels tend to be higher in older people. The average woman in the 16-49 age range is older than the average pregnant woman. Failing to account for the fact that women who actually get pregnant are younger than the average woman of "childbearing age" caused EPA to greatly overstate the fraction of children born to women with mercury exposures exceeding the RfD.

Doubling the estimate due to higher mercury levels in umbilical cord blood was double counting, because EPA already accounted for this when it set the RfD. To set the RfD, EPA begins with the lowest mercury level associated with any health effect in epidemiological studies. In this case, the starting point was a level in maternal blood of 85 ppb.[189] EPA then takes the bottom of the 95 percent confidence interval for this estimate, which happened to be 58 ppb.[190] Then EPA divides this number by safety factors to make sure that no one comes anywhere near an exposure level that might cause harm. In this case, a safety factor of 3.15 was included specifically to account for the fact that mercury is more concentrated in umbilical cord blood.[191] Overall, EPA included a factor of ten for safety, setting the RfD at 5.8 ppb, or one-tenth of 58 ppb.

In other words, EPA doubled its estimate of how many babies are exposed to mercury at a level greater than the RfD based on a factor that had already been accounted for in the original estimate.

Recent data suggest mercury exposures are even lower than suggested by the data collected during 1999–2000. Pregnant women's mercury levels were nearly 50 percent lower in the NHANES data collected during 2001–2 and 2003–4 when compared with pregnant women measured during 1999–2000.[192]

Putting this all together, and using data on nearly 900 pregnant women collected from 1999 to 2004, the fraction of pregnant women with mercury above the RfD was 2.4 percent, rather than the 16 percent claimed by EPA. Using the 2001–4 data alone, the fraction was less than 0.4 percent.[193]

It isn't clear whether the drop from 1999–2000 to 2001–4 represents a real decline in pregnant women's mercury exposure, sampling bias, or random sampling error. What is significant is that, as of this writing, the lower mercury levels in 2001–2 and 2003-4 have not been reported by any government environmental officials, by environmentalists, or by the media.

In any case, based on data for pregnant women collected from 1999–2004, it appears that EPA overstated by nearly a factor of seven the number of children born each year to mothers whose mercury levels exceed the RfD.

NHANES also measured mercury levels in children. These data likewise show that few children are exposed to mercury at levels above the RfD. Among nearly 2,500 children up to five years of age measured from 1999 to 2004, only 0.7 percent exceeded the RfD. The highest mercury level in any child was 10.4 ppb, or less than twice the RfD.[194] Once again, this shows that exposure to mercury in excess of the RfD is rare, rather than common, as EPA and its allies claim.

Second, regulators and activists have also overstated the health risk from any given level of mercury exposure. Even the exposure of only 1 or 2 percent of babies to mercury above the RfD would still be cause for concern if exceeding the RfD really did put them at risk of mental retardation. But this claim is in the realm of hysterical fantasy. The highest mercury exposure measured in a pregnant woman in NHANES was 21.4 ppb during 1999–2004, or 3.7 times the RfD. Of pregnant women who exceeded the RfD, 75 percent were below two times the RfD. There is no evidence for even the mildest health effects at mercury exposures this low.

The RfD is set at one-fifteenth the level that was associated with the mildest health effects measured in a study of children in the Faroe Islands—that is, one-fifteenth of 85 ppb, as described above.[195] Children in the Faroe Islands are exposed to much higher mercury levels than Americans, mainly through consumption of whale meat.[196] Many in the Faroes study had mercury levels well above *ten or fifteen times the RfD*, yet none showed signs of mental retardation or learning disabilities. The health effect used to set the RfD involved small reductions in scores on the Boston Naming Test (BNT), a test in which children make line drawings of objects.[197] EPA chose the BNT results because they appeared to be the most sensitive to mercury exposure among the twenty tests administered to the children in the Faroe Islands study.

Associations of mercury with reductions in performance on other specific tests (such as finger-tapping or recall of names) or on more general tests of cognitive performance either required higher mercury exposures or did not occur at any exposure level. Mental retardation and other serious health effects were not observed in even the most highly exposed children in the study (whose exposures were still substantially lower than those associated with the serious poisoning events in Japan and Iraq).

In other words, in the Faroe Islands study, even at exposures fifteen times greater than EPA's reference dose and several times greater than even the highest levels measured by the CDC in pregnant U.S. women, mercury was associated only with mild and clinically unnoticeable health effects.

Children in the Faroe Islands study were not even given an IQ test (though they took three of thirteen subtests of the WISC-III IQ test, as well as several other more specific cognitive performance tests). Two other studies that did administer IQ tests to children with high mercury exposures, however, did not find any reduction in IQ at any mercury exposure level. For example, a study of children in the Seychelles administered forty-eight different neurological and cognitive tests, including an IQ test, and found no harm from mercury, even though exposure levels were similar (actually somewhat higher) to those in the Faroe Islands.[198] A study of New Zealand children exposed to mercury from fish likewise did not find any change in IQ associated with mercury.[199]

The Seychelles and New Zealand studies are more relevant to people in the United States because the children in them received their mercury through fish rather than whale meat, as was the case in the Faroes. The whale meat eaten by the Faroese also has about five times the mercury per unit mass as the fish eaten by the Seychellois.[200] Thus, to the extent mercury is actually causing neurological deficits in the Faroese, it could be due to higher acute exposures from whale meat, when compared with exposure through fish. The Seychellois are also ethnically more representative of Americans, being descended from white Europeans and black Africans, while the Faroese are homogeneously Scandinavian.[201]

Earlier in this chapter we discussed the ubiquitous problem of spurious risk estimates in observational epidemiologic studies. The Faroe Islands study is no exception. The study reported that "occasional alcohol intake during pregnancy was unexpectedly associated with improved performance in the children."[202] The authors attribute this result to demographic differences between pregnant women who lived in the capital city of Tòrshavn versus other areas of the Faroe Islands. Tòrshavn women were more likely to consume alcohol during pregnancy. Both they and their children also had higher cognitive performance scores than families in other areas of the Faroe Islands. But they also had lower mercury exposures than people in other areas, because people in the capital city consume less whale meat than other Faroese.

In other words, residence in the capital city was associated with higher alcohol consumption, lower mercury exposure, and higher cognitive test scores. Nevertheless, the researchers did not control for capital city residence in their analysis of mercury's cognitive effects. The study therefore likely suffers from confounding based on whether a pregnant woman resided in the capital city. The cognitive differences that the researchers attributed to mercury instead could have resulted simply from demographic differences between people in the capital city versus other areas of the Faroe Islands.

Despite the evidence that mercury in fish is not harmful even at doses many times greater than Americans are ever exposed to, media and activist accounts continue to imply or even explicitly claim that hundreds of thousands of American children are suffering serious brain damage from mercury in fish. Here are typical examples:

- From the *Atlanta Journal-Constitution*: "When mercury is consumed by a pregnant woman, most often when she eats fish, it can cause her baby to be born with brain damage. Although the effect can be severe in individual cases, a report by the National Academy of Sciences warned in 2000 that mercury poisoning of unborn babies in America probably results in an overall increase in the number of children 'who have to struggle to keep up in school.' The EPA has estimated that each year 630,000 newborns in the United States, or nearly one in six, have dangerous levels of mercury in their blood."[203]

- From the *Dallas Morning News*: "Ingested in sufficient quantities, mercury—a byproduct of coal combustion—can harm the nervous system and cause learning disabilities, mental retardation and other problems. It's a particular threat to fetuses exposed through their mothers; the EPA estimates that 630,000 of the 4 million babies born each year could be at risk for some type of mercury-related developmental disorder."[204]

- From the Associated Press: "Mercury exposure can cause permanent brain and kidney damage, said Dr. John Pittman, and unborn and young children are particularly at risk. The EPA estimates that as many as 630,000 children may be born each year with unhealthy levels of mercury in their blood. 'The amount of mercury (in patients) is through the roof,' Pittman said."[205]

- According to the Natural Resources Defense Council, "Eating fish contaminated with mercury, a poison that interferes with the brain and nervous system, can cause serious health problems, especially for children and pregnant women," and "Prenatal and infant mercury exposure can cause mental retardation, cerebral palsy, deafness and blindness."[206]

- A Friends of the Earth press release highlights "a hard-hitting national ad in today's national edition of *USA Today* . . . [that] shows an image of toddlers with the headline 'They're being poisoned.'"[207]

- A *USA Today* story on mercury in fish warned that "some fish are mercury-filled time bombs, according to a parade of reports from government agencies and environmental groups."[208]

These claims demonstrate the degree to which public understanding about mercury has become almost completely disconnected from reality. Even in the Faroe Islands study, none of the children were learning-disabled or mentally retarded. The effects associated with mercury, even if assumed to be real, would be undetectable even to teachers or psychologists. The only way the Faroe Islands researchers were able to report any apparent harm from mercury at all was by administering a large number of specialized cognitive and neurological tests and finding small, clinically unnoticeable, decreases in average scores on a few of them.

Conclusion: Regulatory Costs and Americans' Health

A sound understanding of the extent to which air pollution affects health might not be so critical if we could reduce pollution for free. If pollution reduction were free, we could put our misgivings about the health-effects research aside and simply implement all available reduction measures, just to be on the safe side. But reducing air pollution is costly. Attaining the federal 8-hour ozone and annual $PM_{2.5}$ standards will cost tens to hundreds of billions of dollars per year.[209] The 8-hour ozone standard might not be attainable at all in some areas of the country.[210] These costs are ultimately paid by people in the form of higher prices, lower wages, and reduced choices.[211] We all have many needs and aspirations, and insufficient resources with which to fulfill them. Spending more on air quality means spending less on other things that improve our health, safety, and quality of life.

Higher incomes are associated with improved health, because people spend a portion of each additional dollar of income on things that directly or indirectly improve health and safety, such as better medical care, more crashworthy cars, and more nutritious food.[212] People made poorer by the costs of regulations do fewer of these things and are less healthy as a result. Risk researchers estimate that every $17 million in regulatory costs induces one additional statistical death.[213] Thus, regulations are not pure risk reduction measures, but instead inevitably impose tradeoffs between the health benefits of the regulation and the harm from the regulation's income-reducing costs. The costs of attaining the ozone and $PM_{2.5}$ EPA adopted in 1997 will likely be more than a thousand dollars per year for each American household.[214] For these huge expenditures, we would eliminate at best a tiny fraction of all disease and disability.

Even if we could somehow convince ourselves that additional air pollution reductions would confer net benefits, focusing on air pollution would still be a foolish policy, because other measures would provide far greater health benefits per dollar invested. Based on an assessment of more than five hundred lifesaving measures in four categories—environmental pollution reduction, workplace safety, injury prevention, and medical care—researchers at the Harvard School of Public Health concluded that environmental measures saved by far the fewest years of life per dollar invested.[215]

We could glibly say that we should undertake all available risk-reduction measures and save as many lives as possible. But this begs the question. If we lived in a world of infinite resources and omniscience about the full consequences of our actions, then of course we would undertake literally all health and safety measures available. But in such a world there would be no politics or policy debates over environmental regulations or over anything else. Politics and policy debates exist exactly because resources and knowledge are scarce and insufficient to satisfy all our needs and aspirations. Maximizing human welfare requires spending these scarce resources in ways that generate the greatest health and welfare improvements per dollar invested. Spending money on air pollution means choosing to save far fewer lives than if the same amount of money were spent in other ways.

One might argue that talking about other ways to reduce risk is irrelevant, because it is not as if money is sitting around waiting to be spent on risk reductions, and air pollution is just one of many choices. We can choose to reduce air pollution or not, but if we choose not to, this does not mean the government will fund some other risk-reduction measure(s). This reasoning implicitly assumes that only publicly determined risk-reduction priorities and expenditures are legitimate. But if people aren't forced to spend money to attain EPA's standards, they will have more money to spend as they see fit. They will spend these funds to improve their health, welfare, and quality of life as they define it. And they will be better off than if they had been forced to spend the money on air pollution reductions that deliver tiny benefits compared to the costs imposed.

In reality, air pollution affects far fewer Americans, far less often, and far less severely than activists, regulators, or health experts would have us believe. We are giving up a great deal to finance regulators' and environmentalists' ever-expanding war on air emissions and getting little in return.

8

Has the Clean Air Act Been
Good for Americans?

We've dramatically reduced air pollution during the last few decades, and air pollution at current levels is at worst a minor risk for Americans' health. Thus, it might be tempting to conclude that the Clean Air Act (CAA) and EPA's regulations and requirements have been a resounding success, and that we just need to do more of what we've been doing. But the fact that pollution has declined since the 1970 passage of the CAA does not tell us whether the Clean Air Act/EPA air quality management system was an especially good way to get there or is now a good way to manage future air quality policy.

At the very least, the Clean Air Act was not *necessary* for continued improvements in air quality. Air pollution had already been dropping for decades before Congress nationalized air pollution regulation in 1970 (see chapter 1). Likewise, environmental quality in general, as well as workplace and road hazards and other risks, was also steadily improving for decades before the federal government seized policy control from state and local governments and ushered in the era of centralized administrative regulation. In each case, the rate of improvement was about the same before and after the federal government stepped in.

After several decades living under what legal scholar David Schoenbrod calls the "modern administrative state," it probably seems unimaginable that "public" goods such as cleaner air or safer cars could be delivered without powerful national agencies commanding that it be so.[1] Unfortunately, we didn't merely trade a decentralized system for an equivalent centralized one. Instead, federal air quality regulation has added a great deal of collateral damage into the bargain. The Clean Air Act creates large administrative burdens, economic distortions, and perverse incentives that impose costs on Americans many times greater than would be necessary merely to reduce air pollution to health-protective levels as efficiently as possible. Indeed, some CAA requirements actually slow progress on air quality or even worsen it.

The collateral damage is only increasing as EPA continues to adopt ever-stricter air pollution limits and ever-tougher regulations to attain them. And there is no end in sight, because the Clean Air Act endows EPA with the power to keep expanding its power. Because EPA sets national air pollution standards, the agency in effect gets to decide when its own job is finished. This conflict of interest goes a long way toward explaining the ubiquitous exaggeration of air pollution levels and risks by regulators and their allies.

Virtually everyone would agree that we all have a right to be free from unreasonable risks imposed by others. But current federal air pollution standards and regulations go well beyond that goal. Rather than a system designed to deliver sufficiently clean air at the least possible cost, America's air quality management system has largely ended up helping special interests—regulators, environmentalists, businesses, and politicians—gain money, power, and prestige and advance their ideological goals at the expense of the American people.

In this chapter, we argue that the Clean Air Act has on balance made Americans worse off and that a more decentralized, results-focused, and legislatively accountable approach would ensure clean air, but with fewer of the harmful side effects of our current system.

A Process-Focused System

If Congress wanted states to achieve a given level of air quality, it could simply have dictated to states (1) the standards and the dates by which those levels would have to be achieved, (2) how attainment would be measured, and (3) the penalties for failure. Given sufficiently large penalties, states would have an incentive to find effective means of meeting their obligations. Such a Clean Air Act could be written on a few pages and would require few or no EPA regulations.

Instead, the Clean Air Act is hundreds of pages long and includes exquisitely detailed requirements for everything from the composition of gasoline to the contents of permits-to-operate for industrial facilities. EPA has written thousands of pages of regulations to implement the CAA requirements, along with tens of thousands of pages of "guidance documents" to explain what the regulations mean. States must develop their own laws and regulations to implement the federal requirements, and businesses must obtain permits that often specify operating conditions and pollution control methods, unit by unit and process by process, and that must be amended any time a production process is changed. These permits are often hundreds of pages long. Legions of lawyers and consultants help businesses figure out what the rules mean and how to comply with them.

The vast majority of this activity has nothing to do with reducing air pollution, but instead involves creating and then demonstrating compliance with administrative requirements. Indeed, just a few emissions limits—mainly for motor vehicles and power plants, which accounted for most of the pollution to begin with—are responsible for the vast majority of air pollution reductions achieved since the Clean Air Act was enacted.

These massive procedural burdens have little to do with improving air quality even though they impose substantial costs for both government agencies and private businesses and individuals who must carry them out. However, they do support large numbers of jobs at state and federal agencies, and the obsessively detailed requirements give regulators and environmental activists tremendous power to micromanage private decisions.

One window into the process-focused nature of air quality regulation is the State Implementation Plan (SIP) process—the centerpiece of the CAA, through which states demonstrate to EPA how they plan to reduce pollution and ultimately attain federal air standards. A SIP includes state and local air pollution officials' inventory of the estimated emissions from all sources in a region, a series of pollution control measures that the region commits to implement to reduce pollution, and an "attainment demonstration"—a combination of computer modeling and other analyses that purports to demonstrate that the region will attain federal pollution standards once the SIP control measures are fully implemented. Once approved by EPA, the control measures and other commitments in the SIP become legally enforceable.

Despite the ostensible goal of improving air quality, in reality it is far more important to have an EPA-approved SIP than actually to reduce air pollution or attain federal air standards. If a state fails to obtain EPA approval for its SIP, EPA can withhold federal highway funds to and limit economic development in areas of the state that violate federal standards. In contrast, if a state fails actually to clean the air or attain federal standards, the main "consequence" is that EPA can, and typically does, extend the attainment deadline and require that the state submit a new SIP to EPA for its review and approval.[2]

SIPs are huge, complex documents that virtually defy comprehension or analysis.[3] Many run to thousands of pages, and more than a thousand state and federal regulators around the country spend most or all of their time creating, maintaining, amending, and reviewing them, while probably thousands more at regulated businesses and activist groups closely follow and attempt to influence their provisions.[4] The *Federal Register*, a daily government publication that reports on all federal regulatory actions, includes hundreds of entries each year detailing EPA reviews, amendments, approvals, and disapprovals of states' SIPs.

The SIP planning process and the regulations that flow from it take place within a world that has only a peripheral relationship to reality. For example, the emissions inventories used in SIPs have been known since at least the late 1980s to be inaccurate and have repeatedly failed real-world validation tests.[5] One key problem is that field studies indicate that gasoline vehicles contribute anywhere from 50 to 75 percent of human-caused VOC emissions in U.S. metropolitan areas, but official inventories typically estimate that only about 40 percent of VOCs come from gasoline vehicles.[6] Nevertheless, these official emissions inventories, which run to hundreds of tons per day for a typical metropolitan area and yet are reported source by source with purported tenth-of-a-ton-per-day precision, are treated as real in the regulatory process and continue to misinform multibillion-dollar pollution control and transportation infrastructure decisions.

Starting from this inaccurate emissions inventory, EPA then prospectively awards emissions-reduction "credit" for pollution control measures in a given SIP based on assumptions about how effective the control measures will be. A state's goal is then to garner enough SIP credit to demonstrate on paper that it will attain the air quality standards. The inaccuracy of the inventory guarantees that SIP credit determinations will be erroneous, but the situation is much worse than that. EPA even awards SIP credit for programs that are known to be ineffective.

For example, EPA awards substantial SIP credit for vehicle emissions inspection and maintenance (I/M) programs, despite a well-documented history of their real-world ineffectiveness.[7] At the same time, EPA does not award any SIP credit for other potential methods of reducing emissions from high-polluting cars, such as using an inexpensive technology called remote sensing to find high polluters as they drive on the road and then requiring that they be repaired or scrapped. In fact, EPA considers it double counting to award SIP credit for on-road remote sensing, because that credit has already been awarded for I/M programs. Since states can receive SIP credit only for traditional, scheduled I/M programs, that is what they implement.

For their part, state regulators also protect and defend the SIP credit system. EPA began requiring I/M programs in many areas during the 1980s, and the state bureaucracies that have grown up around these programs have an interest in their continuation. The perverse result is that even as on-road measurements continue to find many high-polluting cars driving out in the real world, regulators implicitly make believe they do not exist, since they have been dealt with—on paper—by I/M programs.

Because states receive SIP credit up front, they have little incentive to perform a valid evaluation of the real-world effectiveness of their programs, and EPA has not required rigorous evaluations. There is no upside for state and federal regulators to such an effort, but there is a tremendous potential downside for both, because a negative evaluation could result in the loss of both the SIP credit necessary to keep the highway funds flowing and the political legitimacy of air pollution control programs. There is also no incentive to seek alternative strategies that would clean the air in the real world, because EPA has decreed up front what "works" for the purposes of meeting Clean Air Act process requirements.

The air quality regulatory system is thus structurally insulated from having to base plans and regulations on realistic estimates of emissions or on the real-world effectiveness of emissions-control measures. In the case of automobile emissions, the result has been billions wasted on ineffective vehicle inspection programs at the same time that both regulators and activists have forgone the huge air quality improvements that could be achieved rapidly and inexpensively through an effective program to find and either repair or scrap high-polluting automobiles.

The discussion above is not merely hypothetical. Data from a wide range of sources show that a small percentage of automobiles accounts for most automobile emissions.[8] As far back as the early 1980s, research sponsored by the California Air Resources Board showed that the worst 10 percent of cars produced about half of all of carbon monoxide emissions from cars.[9] Recent on-road data collected via remote sensing indicate that now the worst 5 percent of cars produce about 50 percent of tailpipe VOC and CO emissions.[10] As the vehicle fleet has become cleaner over time, fewer cars become high polluters, and the emissions distribution has become even more skewed toward a few very high-polluting cars. Figure 8-1 shows the distribution of VOC emissions from more than 1,200 vehicles measured twice by on-road remote sensing in Southern California in 2001. Note that most have hardly any VOC emissions, while a few have enormous emissions.

This heavily skewed distribution suggests that (1) repairing or scrapping a few percent of cars could reduce VOC emissions from the automobile fleet by as much as 30–40 percent, and (2) most emissions testing in scheduled I/M programs involves testing clean cars. High-polluting cars continue to be found on the road in spite of the existence of I/M programs, and researchers have found that many of these cars either avoided or passed their scheduled I/M test despite their high emissions.[11] Thus, I/M programs are both ineffective and inefficient.

Since at least the early 1990s, scientists and policy analysts have pointed out that an on-road measurement program to identify and ensure repair or retirement of high-polluting vehicles could generate large emissions reductions.[12] Nevertheless, there is little constituency for such a program among regulators or environmental activists, who continue to support traditional I/M programs. Americans continue to spend more than a billion dollars per year in traditional vehicle inspection programs, and the large and inexpensive pollution reductions potentially available from an on-road program targeting high polluters remain unclaimed.

FIGURE 8-1
VOC EMISSIONS DISTRIBUTION OF VEHICLES MEASURED BY REMOTE
SENSING IN RIVERSIDE, CALIFORNIA, IN 2001

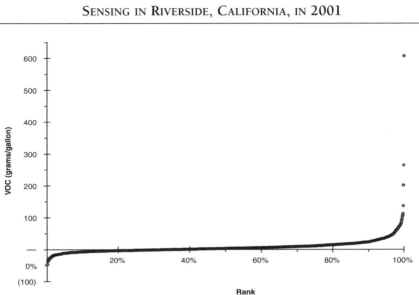

SOURCE: Data were collected in Riverside, California, in 2001 by Gary Bishop of the University of Denver. Data were downloaded from Fuel Efficiency Automobile Test Data Center, Light-Duty Vehicles, www.feat.biochem.du.edu/light_duty_vehicles.html (accessed February 23, 2003).

NOTES: VOC emissions from 1,207 vehicles with two or more measurements, ranked from cleanest to dirtiest. These vehicles are fifteen years old on average, and 75 percent are at least twelve years old. Emissions are in grams of VOCs emitted per gallon of fuel burned. Negative VOC readings do not represent negative emissions (which is physically impossible), but all measurements include some random noise, resulting in some vehicles with very low or zero VOC emissions registering as having negative emissions.

Clean Air Act requirements for large industrial sources provide another example of how regulation can unnecessarily increase costs while at the same time actually slowing the pace of air quality improvement. New Source Review (NSR) is a key component of the Clean Air Act's provisions for regulating industrial air pollution. NSR requires businesses to install "state-of-the-art" pollution controls when they build new plants or make major modifications to existing ones. For example, NSR requires facilities in areas that violate EPA standards to install pollution controls to meet the Lowest Achievable Emissions Rate (LAER), as determined by EPA, without regard to the cost of controls.

New Source Review has a number of perverse effects.[13] First, it makes new and upgraded facilities relatively more expensive than existing ones, encouraging businesses to put their research and development funds into finding ways to keep older, less-efficient, and higher-polluting plants operating well beyond their nominal useful lives. Second, all new or modified facilities are required to install state-of-the-art pollution controls, even if they are already relatively low-emitting. NSR thus funnels resources into comparatively high-cost/low-benefit pollution reductions.

Power plants tend to be long-lived and are therefore a prime example of the perverse incentives created by New Source Review. Older coal-fired power plants emit NOx at rates of 0.65–1.50 pounds per million British Thermal Units (lb/mmBTU), a measure of NOx emitted per unit of fuel burned.[14] Some of these plants could reduce NOx for as little as $300 per ton of pollution eliminated. In contrast, new natural-gas-fired power plants without any add-on pollution controls emit about 0.05–0.10 lb/mmBTU, or 85–97 percent less than the old coal-fired plants. Nevertheless, NSR requires new gas plants to meet LAER—about 0.02 lb/mmBTU—at a cost of $2,500 to more than $10,000 per ton—as much as eight to thirty-three times more than the cost of reducing the same amount of pollution from old coal-fired plants.[15]

NSR, in effect, requires the most expensive and inefficient pollution reductions and greatly increases the costs of building new, more efficient power plants. The predictable result is that NSR has encouraged the continued operation of older coal-fired plants and has therefore caused higher pollution levels than might have occurred under a regulatory system that treated old and new facilities the same (more on this below).

NSR also does not create an ongoing incentive to reduce pollution. For example, a facility that goes through NSR must meet the emission-rate standards that apply at the time, but need not reduce pollution any further in the future unless it makes a modification that triggers NSR again. Finally, NSR imposes additional "transaction costs" through EPA's detailed micromanagement of the process. It can take as much as a year and a half for a company to get a permit to proceed with construction under NSR.[16]

If this seems somewhat abstract, perhaps an analogy with everyday life would be helpful. Imagine you have an old house, and your old, rusty, inefficient water heater breaks down. You could repair it, but most people would find it more cost-effective and beneficial to replace it with a new, probably larger, and more energy efficient water heater.

But imagine if you had to go through the equivalent of New Source Review when your water heater breaks. If NSR applied to your house, you couldn't install a new water heater without also modernizing and upgrading your entire home. You'd have to replace your single-paned windows with double-paned windows, put insulation in all your walls, add a sprinkler system, replace galvanized pipes with copper, replace your old electrical outlets with modern ground-fault indicator outlets, and so on. If we applied NSR to houses, few people would upgrade or replace anything in their houses unless they simply couldn't patch up a piece of essential equipment. Our houses would be less comfortable, less energy-efficient, and less safe as a result. NSR has the same effects.

Despite its deficiencies, NSR is a sacred cow for regulators and environmentalists, who have vigorously defended it against even modest attempts at reform. Robert Kennedy Jr. of the Natural Resources Defense Council has called NSR "the heart and soul of the Clean Air Act."[17] Even businesses have sought only to soften the portion of NSR that applies to existing facilities. None of the major "stakeholders" have tried to eliminate NSR and replace it with pollution taxes or cap-and-trade systems, even though either approach would provide greater certainty about when and by how much emissions would be reduced and would allow more pollution to be eliminated for any given expenditure.

The persistence of NSR in spite of its harm to air quality becomes more understandable when we consider its benefits to stakeholders in the regulatory process. NSR gives regulators a great deal of power to micromanage private decisions and, like administratively complex regulations in general, creates a need for more regulators and larger agency budgets to carry out the administrative duties. For environmentalists, NSR's procedural requirements create many administrative decision points at which they can intervene, generate media publicity, and attempt to delay or prevent businesses from carrying out their plans or force alterations in those plans along the lines environmentalists desire. Even businesses often find NSR beneficial on balance, because it protects them from competition by imposing large costs on potential competitors, discouraging new entrants to the market. Streamlined systems such as cap-and-trade or pollution taxes don't offer these benefits.

The Clean Air Act includes other inflexible requirements that misdirect resources and encourage wasteful activities. For example, parts of California's San Joaquin Valley (SJV) violate federal PM_{10} standards. But in the SJV, high PM_{10} levels occur almost exclusively during the winter. Furthermore, only two or three out of more than a dozen monitoring locations violate the standards. Nevertheless, the Clean Air Act requires the SJV to implement Best Available Control Measures (a Clean Air Act term of art) for PM_{10} year-round and everywhere in the region.

The CAA's "rate of progress" (ROP) provision requires ozone nonattainment areas to demonstrate an average reduction in ozone precursors (VOCs and NOx) of at least 3 percent per year. In addition to this demonstration being based on a spurious emissions inventory and paper SIP credit that often have no real-world validity (see above), the ROP provision encourages regions to seek pollution reductions wherever they can be obtained, rather than in specific areas that most need them to progress toward attainment of federal air standards.

The SJV once again provides an example. The northern areas of the SJV (the Stockton and Modesto areas) already attain the 8-hour ozone standard, while the eastern areas of the central and southern SJV (the Fresno and Bakersfield areas) exceed the standard by substantial margins. Nevertheless, reductions *anywhere* in the SJV count for meeting the ROP requirement. Since it is often easier to achieve modest reductions in each of several areas than large reductions in a few areas, the ROP requirement has encouraged the SJV to require emissions reductions in areas of the SJV that have little or no ozone problem, while devoting too little effort to achieving reductions where they are most needed for compliance with federal ambient air quality standards.

The issuing of permits is perhaps the most process-oriented feature of air pollution control. Rather than simply limiting the total emissions from a given facility, industry, or activity, EPA's regulations and guidance spell out a range of requirements in excruciating detail, much of it incorporated into the operating permits that businesses must obtain to comply with the Clean Air Act. David Schoenbrod, a law professor and former litigator for the Natural Resources Defense Council, describes the permitting system this way:

> Instead of limiting total emissions from each plant, the regulatory system frequently slaps a separate emissions limit on every

one of the many smokestacks, pipes and vents coming out of the typical plant. [EPA] regulates not only emissions but sometimes also the techniques used to control them, monitor them, and report them. All this must be pinned down in a permit. . . . If the source needs to change what it is producing or how it operates . . . it will need an amended permit. Beyond all this, a source that violates a requirement can be punished heavily even if no harm was done and even if the violation was neither intentional nor negligent. . . . No major facility can hope to avoid violating such an exacting system of legally binding requirements. . . . More than a few former colleagues of mine at the Natural Resources Defense Council, who now work for corporations trying to comply with environmental law, tell me . . . their clients can't help but violate the law, no matter how hard they try, because the legal requirements are just too complex and confusing.[18]

Whatever success the nation has achieved in improving air quality, it has and continues to come at a far higher cost than necessary, not only in terms of direct costs, but also in terms of administrative overhead and intrusion of government bureaucrats and environmental activists into a growing range of formerly private decisions.

Environmentalists, Regulators, and Other Special Interests

No one intent on improving air quality would set out to design a system that wastes most resources on process, ineffective programs, and unnecessarily expensive measures and that has many harmful spinoff effects. And yet legislators, regulators, businesses, and environmental activists have created such a system and allowed it to persist and expand over the thirty-seven years since the federal government nationalized air pollution policy in 1970. How can this outcome be explained?

In the conventional view of environmental issues, businesses and their lobbyists are the special interests who want to profit at the expense of people and the environment. Environmentalists stand up to these special interests on behalf of the public. Regulators fall somewhere in between, depending on the particular agency and the specific regulatory issue.

This is a naïve view of environmental regulation. A key to understanding the structure of air pollution regulation is to realize that, like businesses, activists, regulators, and legislators have private interests of their own that may depart substantially from the public's interest in attaining sufficiently clean air at the lowest possible cost. These include money, power, job security, institutional growth, ideology, and prestige.

A number of analysts have shown how a complex, centrally controlled, bureaucratic, and process-focused regulatory system actually serves the interests of environmental activists, legislators, federal and state regulators, and even many regulated businesses.[19] For many readers, this conclusion may seem counterintuitive. As Jonathan Adler, a legal scholar at Case-Western Reserve University, has noted:

> Most Americans recognize that politics has a lot to do with the pursuit of power, privilege, and special interests; however, there is a general presumption that environmental politics is somehow different. We take for granted that environmental laws are what they seem; that the legislators who enact those laws and the bureaucrats who implement them are earnestly struggling to protect public interests; and that these laws will be enforced in a fair and sensible manner. All too often, however, environmental regulations are designed to serve narrow political and economic interests, not the public interest.[20]

Nevertheless, an expanded concept of "special interests" helps explain why air quality regulation is not primarily focused on real-world benefits for the people it supposedly protects.

While businesses are often portrayed as opposing government regulations, businesses often *seek* government regulations. Appropriately structured regulations can provide businesses with direct financial benefits, restrict competition, or harm competitors. Here are some examples:

- The 1990 Clean Air Act Amendments required "oxygenates" to be added to gasoline, with the ostensible goal of reducing carbon monoxide and ozone levels. Ethanol, a key oxygenate, is produced from corn grown in the Midwest. Midwestern farmers and agribusinesses lobbied heavily to create a guaranteed market for ethanol in the Clean Air Act, and midwestern legislators delivered.[21] Despite evidence that oxygenated gasoline actually increases ozone-forming emissions, EPA refused to grant states' petitions for waivers of the oxygenate requirement.[22] In the 2005 federal energy bill, Congress did eliminate the oxygenate requirement, but replaced it with a guaranteed market for ethanol producers.[23] American gasoline must now include at least 4 million gallons of ethanol in 2006, with the mandate rising to 7.5 billion gallons in 2012.[24] Ethanol producers currently receive a federal taxpayer subsidy of fifty-one cents for every gallon of ethanol blended into gasoline, and are protected from cheaper ethanol from foreign producers by a 54-cent-per-gallon tariff on imported ethanol.[25] Ethanol mandates continue to funnel billions of dollars per year from motorists to politically powerful agribusinesses.

- New Source Review, discussed earlier for its role in impeding environmental progress, also helps existing businesses by imposing large additional costs on new facilities, putting those new facilities at a competitive disadvantage. Thus, NSR not only encourages businesses to keep older plants running; it protects those plants from competition by new and potentially more efficient upstarts. This harms consumers as well as air quality by raising prices, slowing innovation, and lowering the quality of goods produced.[26]

- One way to reduce sulfur dioxide emissions from coal-fired power plants is to switch to low-sulfur coal. In fact, this is how many eastern coal-fired power plants met the 1970 Clean Air Act's new source performance standard for sulfur dioxide emissions. But most low-sulfur coal is produced in the western United States, while midwestern and eastern coal has a relatively high sulfur content. Thus, the switch to western coal harmed eastern coal producers. Rather than attempt to weaken the regulation of sulfur dioxide emissions from coal, eastern high-sulfur coal producers instead lobbied successfully to require all new coal plants to install sulfur scrubbers—the most expensive method of reducing SO_2 emissions—no matter how clean their coal was. Environmentalists also lobbied for the scrubbing requirement, which was included in the 1977 Clean Air Act Amendments.[27] This eliminated the rationale for using low-sulfur coal, since it was costly to transport from west to east, and the new standard could be met by scrubbing alone. Midwest and eastern high-sulfur coal business owners and miners also lobbied for the scrubber requirement, because it protected the market for their high-sulfur coal.[28] Ironically, the scrubbing requirement extended the life of older, dirtier coal plants by making them more competitive relative to newer facilities, and also greatly increased the amount of scrubber sludge requiring disposal.[29]

Environmental activists gain power, prestige, public relations opportunities, and increased funding in a centralized, complex, and politicized regulatory system with many administrative decision points. Such a system allows environmental lobbies to play a prominent role in national policy, both in drafting federal legislation and in using litigation to pursue policies not enacted through the legislative process.[30] These activities also create public relations opportunities that help with fundraising.

Furthermore, the many procedural requirements of environmental statutes create opportunities for citizen suits for technical violations of environmental statutes without the need to show that anyone was actually harmed. This can also be a profit-making activity, because activists are reimbursed for their legal costs based on the going rate for private attorneys, rather than on the basis of their actual litigation costs. These funds can then be used to fund additional suits or other programs. Research on activists' choices of litigation targets "suggests that the decision by these groups regarding the allocation of their litigation resources is driven more by the cost of the action, the ease of victory, and the likely payoff, rather than by the severity of the [environmental] harm or an absence of public enforcement."[31]

Another clue that environmentalists are following the money rather than environmental improvement or public health is that the suits are almost always against private defendants, even though municipalities are more likely to violate their environmental permits.[32] Settlements in these suits often include the establishment of "mitigation" programs, such as environmental education programs, the funds for which often go to environmental groups.

In a recent study of environmental citizen suits, economist Bruce Benson of Florida State University found that lawsuits were filed mainly under those statutes and legal provisions that provided the greatest financial return to the group pursuing the litigation, even though the litigation was often over technical violations that caused no environmental harm, such as violations of paper reporting requirements.[33]

A review of ninety-seven settlement agreements showed that payments to environmental groups who brought the suits averaged more than $97,000 per case.[34] These settlement payments were in addition to revenues generated from attorneys' fees. In order to avoid the appearance of inappropriately profiting from their litigation, environmental groups have even set up "shell" organizations to receive the settlement money.[35]

Money also figures in activists' calculations of what policies to support. A number of researchers have revealed how, despite their ostensible goal of protecting the environment, environmental groups have received funding from businesses and government agencies to support regulations that are ineffective or even harmful to the environment. Coal scrubbing is one example that we've already discussed. As another example, several environmental groups have received funding from the hazardous waste treatment industry. Environmental groups promote regulations that increase the amount of waste classified as hazardous, but that at the same time discourage pollution prevention or recycling. In one case, environmentalists joined the hazardous waste treatment industry in opposing a tax that would have been used to fund Superfund cleanups, but would also have encouraged industry to reduce the amount of hazardous waste it produces.[36]

EPA and vehicle emissions–testing contractors have funded the American Lung Association to campaign for vehicle emissions inspection programs, and ALA has lobbied for expansion of I/M.[37] Yet, as discussed above, I/M programs are ineffective and inefficient, while more promising measures to deal with high-polluting cars languish for lack of political support from environmentalists and regulators. The same applies to environmentalists' support for scrubbers and New Source Review.

Regulators also benefit from the command-and-control system. The huge administrative burdens created by the Clean Air Act—the continual generation, review, and amendment of detailed plans, permits, and regulations—ensure a continued need for large numbers of state and federal regulators and give those regulators great power over Americans' lives. Thus, we should expect regulators to favor the complex, bureaucratic systems that give them more power and control and require larger agencies to administer them.

Activists and regulators also have an incentive to maintain a high degree of fear and anxiety over purported environmental risks. If most people believed the air was reasonably safe to breathe, public support for environmental regulations would diminish, threatening activists' and regulators' power and jobs.[38] We've shown throughout this book that activists and regulators exaggerate air pollution levels and risks and obscure positive trends. Such behavior would be inexplicable for organizations that were merely pursuing the public's interest in safe air and the lowest possible cleanup costs. But viewing regulators and activists as just two more self-interested groups provides a more plausible explanation of why they often exaggerate air pollution and risks. They need to keep people scared, regardless of how safe the air is. Thus we have the paradoxical situation in which public fear and pessimism about air pollution has increased even as air quality has continually improved.

The special-interest paradigm also explains the opposition of regulators, activists, and, often, regulated businesses to the use of more flexible, decentralized, results-focused methods of pollution control, such as cap-and-trade programs. Under a cap-and-trade program, total emissions in a region or from an industry are capped at a maximum level that declines each year, and each facility is allocated a portion of the total allowable emissions. To meet the declining cap, a facility can reduce emissions itself or buy emission permits from other businesses that achieved excess emission reductions. Overall pollution declines as required by the cap, but control costs are lower, because each facility has the flexibility to find the cheapest ways to reduce pollution.

The problem for regulators and activists is that in a tradable permit system, the key political decision—how much to reduce pollution—is made up front and covers a wide range of facilities. From then on such programs are "largely self-executing and self-enforcing, resting on decentralized, market-based, decision-making processes."[39] "Market-based" programs like cap-and-trade thereby reduce the power of and need for professional environmental activists and regulators and make their accumulated expertise in the current system obsolete. They also reduce ongoing opportunities to confer regulatory largesse on favored businesses or industries, explaining the frequent lack of support by industry for market-based programs as well.[40] Finally, market-based programs greatly reduce the public relations opportunities inherent in systems like New Source Review, which have many administrative decision points and opportunities for lawsuits and micromanagement. Instead, under cap-and-trade, pollution control occurs quietly and behind the scenes, without micromanagement or political meddling.

War without End

Perhaps the most pernicious feature of the federal administrative state is that it has no negative feedbacks that would slow down or stop its bureaucratic expansion. Just the opposite is true, in fact. All the feedbacks are positive and the system is fraught with conflicts of interest. EPA and state regulators depend on having a serious and urgent problem to solve. But regulators are also major funders of the health research intended to demonstrate the need for more regulation.[41] Regulators decide what questions are asked, which scientists are funded to answer them, and how the results are portrayed in official reports. Thus, environmental health research is not merely a dispassionate scientific enterprise, but is funded with the goal of maintaining and augmenting public anxiety over air pollution. Regulators also provide millions of dollars a year to environmental groups, who then foment public fear of air pollution and agitate for increases in regulators' powers.[42]

Scientific and medical research nominally has more checks and balances than more explicitly political activities, but environmental health research suffers from its own set of pressures to exaggerate risks. As we've seen, studies that report harm from air pollution are more likely to be published and to receive press coverage than studies that do not. Government officials fund much of the research, and the funding is provided with the explicit intent to provide proof of harm from air pollution. Researchers who believe low-level air pollution is a serious threat and who report larger health effects are probably more likely to attract this research funding. Scientists who choose a career in air pollution health research are also probably more likely to hold an environmentalist ideology and to believe that air pollution is a serious problem. Indeed, many environmental health researchers have explicitly associated themselves with environmental groups and causes.[43]

Regulators themselves also create fear through their regional air pollution alert systems. These are the "code red" days and "spare the air" days that regulators declare when they predict air pollution will exceed federal standards on a given day. This constant stream of air pollution warnings maintains anxiety that air pollution is causing great harm. And as the standards are tightened, the number of warnings actually increases, creating a mistaken appearance of increasing air pollution, even as actual air pollution has declined.

The Clean Air Act charges EPA with setting air pollution health standards. But this means federal regulators are the ones who decide when their own jobs are finished. Not surprisingly, EPA has never declared the air safe and continues to tighten the standards to whatever extent is politically feasible at any given time. Congress also charges EPA with evaluating the effectiveness of its own programs. EPA is like a company that gets to decide how much of its products customers must buy and to audit its own books.

Much of this book has shown how regulators, activists, and scientists generally provide the public and journalists with incorrect information on air pollution levels, trends, and health risks. The incentives built in to the Clean Air Act to keep people scared go a long way toward explaining this behavior.

There are other ways that regulatory agencies have missions and goals that are often at odds with the interests of the people they are supposedly protecting. For example, the Clean Air Act Amendments of 1990 and the Intermodel Surface Transportation and Efficiency Act integrated air quality considerations into regional transportation planning via Metropolitan Planning Organizations (MPOs). These are the regional councils of governments that draw up transportation plans for the nation's metropolitan areas. Yet rather than a means to improve air quality, this policy linkage has largely been a pretext for implementing national anti-mobility, anti-suburb policies that are at odds with Americans' actual lifestyle and travel preferences.

In fact, many activists, planners, and regional transportation plans have the explicit goal of increasing road congestion in order to make driving less convenient and encourage people to use public transit.[44] As with other aspects of Clean Air Act regulation, EPA also funds outside organizations to help carry out these anti-mobility efforts and to lobby for greater regulatory powers.[45] Americans use automobiles for about 88 percent of all travel.[46] Efficient automobility is crucial to people's prosperity and quality of life. And as shown earlier, it has long been clear that technology in the form of inherently clean automobiles is mitigating transportation-related air pollution. Yet rather than focusing transportation policy on improving transportation, the Clean Air Act "arguably made air quality the premier objective of the nation's surface transportation programs."[47]

Despite the fact that EPA's 1997 air pollution standards are already more than stringent enough to protect people's health with room to spare, EPA recently tightened the $PM_{2.5}$ standard still further and proposed a tougher ozone standard as well. This new standard will place most of the nation out of attainment for ozone. Furthermore, it will likely be unattainable in many cities, even if virtually all human-caused ozone-forming emissions are eliminated.[48] The new standard will thus make the war on air pollution, and its associated costs, a permanent fixture in America's metropolitan areas. Yet these costs will be borne without any offsetting health improvements to soften their blow.

Better Ways to Achieve Cleaner Air

Before the modern administrative state, decentralized actions were delivering the improved air quality and other health and safety amenities that an increasingly wealthy and educated polity was demanding. Government certainly played a role. But before the era of compulsively detailed administrative regulation, the government's role was, to paraphrase University of Chicago economist Sam Peltzman, complementary to market forces, evolving gradually and incrementally, and largely working in concert with people's values and preferences.[49]

In contrast, today's federal regulatory system imposes revolutionary institutional changes that override people's preferences and suppress individual initiative and creativity. Law professor David Schoenbrod began his career as an idealistic attorney for the Natural Resources Defense Council during the 1970s. But his experiences convinced him that idealistic regulatory systems are unkind to the people they are supposedly protecting. "What the [federal] administrative state added was comprehensive, command-and-control management from on high. The result is an intrusive, inflexible system that bureaucratizes all life that it touches, yet has left the public more anxious about pollution than ever. Such anxiety fuels the growing power of the administrative state."[50]

Greater use of flexible and decentralized approaches and elimination of counterproductive programs like New Source Review would reduce much of the waste in the current system. Even so, this would do nothing to address the underlying problems created by the complexity of the regulations and the incentives for continued regulatory expansion and unwarranted alarmism.

Is there a better way? Schoenbrod has suggested the key features of an admittedly radically reformed air pollution regulatory system: First, the federal government would return air pollution regulation to the states, except for those few cases, such as interstate air pollution, that state governments do not have appropriate incentives or ability to address. Before the 1970 Clean Air Act, state and local governments and the courts were actively and effectively addressing air pollution problems as they were understood at that time. This recommendation would require air pollution policy to be made by legislators closer to the people affected by the laws they enact.

Second, legislation at any level of government would be based on proscribing unjust conduct, such as harming someone's health or property with pollution, rather than on abstract ideals, such as to "protect public health with an adequate margin of safety," as the Clean Air Act requires. Furthermore, the rules would be made by elected legislators, rather than by bureaucrats at administrative agencies. In other words, neither Congress nor other legislative bodies could delegate legislative authority to an administrative agency like EPA. Says Schoenbrod, "With legislators having to take responsibility for the hard choices and deprived of the political profit that comes from unfunded mandates, from blame shifting through delegation, and from case work, legislators would lose their stake in growing the power of the state."[51]

Finally, the judicial process would look more like the common law. For example, rather than giving third parties standing to sue, real injury would be required for standing in court. Legal remedies would, where possible, emphasize compensation for or injunctions against real harm. This would contrast with the current permit-based command-and-control system in which criminal and civil penalties are imposed merely for violating administrative requirements even when there is no negligence and no one is harmed.

Implementing these recommendations would create an air pollution regulatory system more likely to impose pollution controls that are proportionate to the problem, that are focused on results rather than administrative process, that deal with real risks and harms rather than abstract ideals, that take account of the overall effects of proposed solutions on constituents, and that are just.

Of course, the interest groups that benefit from the current system will continue vigorously to defend and promote it under the guise of environmental and public-health ideals. Most Americans are unaware of the costs of the current system or the fact that regulators and environmentalists routinely exaggerate risks. Thus, there is little impetus for change. Worse yet, despite vast improvements in air quality during the last few decades, people are more worried than ever about the ostensible risks of air pollution.

Conclusion

Americans began reducing air pollution as soon as each given pollutant came to be generally recognized as a "problem" that needed to be solved. As a result, pollution declined throughout the twentieth century. Before the Clean Air Act these declines occurred due to market forces and technological advancement, common-law nuisance suits, and *ad hoc* local and state regulation. Improvements continued after 1970 when the federal government took over air pollution policy from the states.

Current air pollution is only a fraction of former levels, and the vast majority of potential health gains available from continued reductions have

already been achieved. Nevertheless, already-adopted measures will eliminate most remaining air pollution during the next twenty years or so.

Yet these environmental successes are filled with irony. Few Americans
are aware of the vast improvement in the nation's air quality. Instead, many
even believe air quality has worsened, will worsen further in the future, and
is still a serious threat to most people's health.

Likewise, few Americans are aware of the wastefulness, intrusiveness,
and inefficiency of our federal air quality management system, and of how
this system serves special interests—environmentalists, regulators, and
businesses—rather than the public interest. Indeed, the general public's
ignorance and unwarranted fear of air pollution are due largely to the trust
it misplaces in two of these three interest groups.

The final irony is that our continued progress on air pollution is a mixed
blessing. While no one would argue that more pollution is better than less,
we are giving up a great deal to achieve each additional increment of pollution reduction, and we will reap at best tiny benefits in return. And so long
as we maintain our bureaucratic, process-focused system of pollution control,
we will pay far too much for the air quality improvements that we do achieve,
and we will continue to give regulators and activists and other government-
empowered busybodies far too much control over how we live our lives. The
war on air pollution will continue to expand so long as EPA gets to set the
standards that keep the agency in business. Changing this state of affairs
will be difficult. The political and regulatory process is controlled by well-
organized groups that have a vested interest in the status quo.

Virtually everyone would agree that we have a right to be free of unreasonable risks imposed by others. But current air pollution standards are
already more than stringent enough to protect people's health. Regulatory
programs are cloaked in the language of public health. But in practice
they mainly protect and expand the powers of federal and state regulators,
create competitive advantage for businesses that can effectively work the
system, and empower environmental activists to override people's preferences and impose their own values regarding how Americans ought to live,
work, and travel.

Americans need and deserve an air quality regulatory system that is
narrowly tailored to solve real problems, rather than used to expand and
perpetuate the power of special interests. A first step to achieving this goal

is more realistic public information about air pollution levels, trends, and, especially, health risks, as well as greater public understanding that regulators and environmentalists often pursue policies that are at odds with the interests and preferences of most Americans.

Journalists have so far failed to turn a critical eye on our air pollution regulatory system or to look beneath the surface of activists', regulators', or health researchers' press releases. Yet among the major providers of public information on the environment, reporters are in the best position to turn the tide of misinformation on air pollution. It would be a breath of fresh air if they took up this challenge. We hope the result will be greater support for a redesigned regulatory system focused on net improvements in Americans' health, welfare, and quality of life.

Appendixes

Appendix to Chapter 2

State-by-State Ozone Trends

Figures 2-6 and 2-7 on pages 204–206 display the average number of 1-hour and 8-hour ozone exceedance days per year from 1985 to 2006, based on all continuously operated monitors in each state. Only states with at least one continuous monitor are included. Note the different vertical scales between the two figures. Table 2-1 on page 208 gives the number of continuous monitors in each state.

We showed above that ozone declined on a national-average basis. These graphs show ozone also declined on a state-by-state basis, or was already low to begin with.

FIGURE 2-6

TREND IN AVERAGE NUMBER OF 1-HOUR OZONE EXCEEDANCE DAYS PER YEAR BY STATE (PLUS THE DISTRICT OF COLUMBIA), 1985–2006

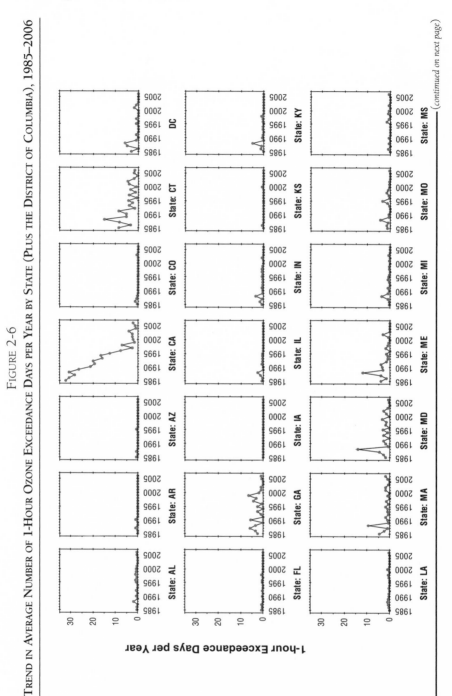

(continued on next page)

(continued from previous page)

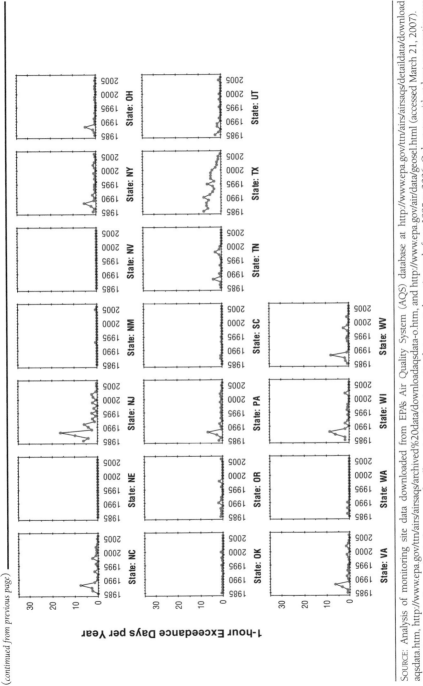

1-hour Exceedance Days per Year

SOURCE: Analysis of monitoring site data downloaded from EPA's Air Quality System (AQS) database at http://www.epa.gov/ttn/airs/airsaqs/detaildata/download aqsdata.htm, http://www.epa.gov/ttn/airs/airsaqs/archived%20data/downloadaqsdata-o.htm, and http://www.epa.gov/air/data/geosel.html (accessed March 21, 2007).
NOTES: Ozone exceedance value is the average for all monitoring sites in a state that operated continuously from 1985 to 2006. Only states with at least one continuous site are included.

Figure 2-7

Trend in Average Number of 8-Hour Ozone Exceedance Days per Year by State (Plus the District of Columbia), 1985–2006

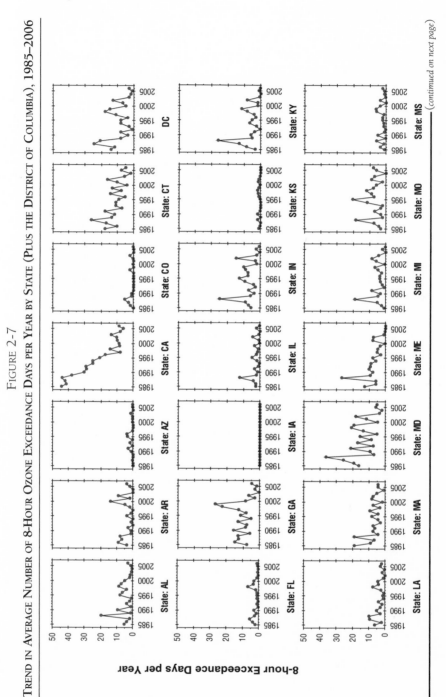

(continued on next page)

(continued from previous page)

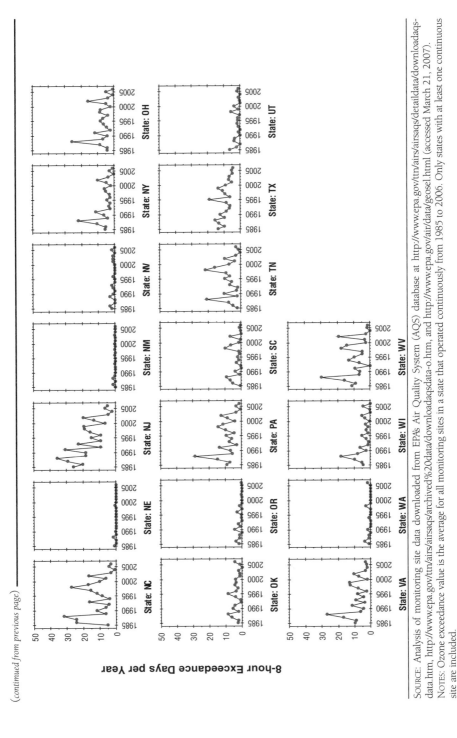

8-hour Exceedance Days per Year

SOURCE: Analysis of monitoring site data downloaded from EPA's Air Quality System (AQS) database at http://www.epa.gov/ttn/airs/airsaqs/detaildata/downloadaqs-data.htm, http://www.epa.gov/ttn/airs/airsaqs/archived%20data/downloadaqsdata-o.htm, and http://www.epa.gov/air/data/geosel.html (accessed March 21, 2007).
NOTES: Ozone exceedance value is the average for all monitoring sites in a state that operated continuously from 1985 to 2006. Only states with at least one continuous site are included.

TABLE 2–1

NUMBER OF CONTINUOUSLY OPERATED OZONE MONITORS
IN EACH STATE (PLUS THE DISTRICT OF COLUMBIA), 1985–2006

State	Number of Monitors	State	Number of Monitors
AL	3	MS	2
AR	1	NC	3
AZ	2	NE	3
CA	41	NJ	9
CO	5	NM	2
CT	7	NV	1
DC	1	NY	8
FL	9	OH	15
GA	2	OK	4
IA	1	OR	2
IL	22	PA	22
IN	8	SC	4
KS	2	TN	7
KY	10	TX	15
LA	3	UT	2
MA	2	VA	7
MD	7	WA	1
ME	2	WI	2
MI	11	WV	2
MO	3	**Nation**	**253**

SOURCE: Analysis of monitoring site data downloaded from EPA's Air Quality System (AQS) database at http://www.epa.gov/ttn/airs/airsaqs/detaildata/downloadaqsdata.htm, http://www.epa.gov/ttn/airs/airsaqs/archived%20data/downloadaqsdata-o.htm, and http://www.epa.gov/air/data/geosel.html (accessed March 21, 2007).

Appendix to Chapter 3

State-by-State PM$_{2.5}$ Trends

Figure 3-7 on the following page gives state-by-state trends in PM$_{2.5}$ levels from 1999 to 2006. The year 1999 was the first time that most states began monitoring PM2.5 as part of the implementation of the new PM$_{2.5}$ standards EPA adopted in 1997. Note that PM$_{2.5}$ declined almost everywhere and that states that started out with the highest PM$_{2.5}$ levels generally achieved the greatest improvements.

FIGURE 3-7

TREND IN ANNUAL-AVERAGE PM$_{2.5}$ LEVELS BY STATE (PLUS THE DISTRICT OF COLUMBIA), 1999–2006

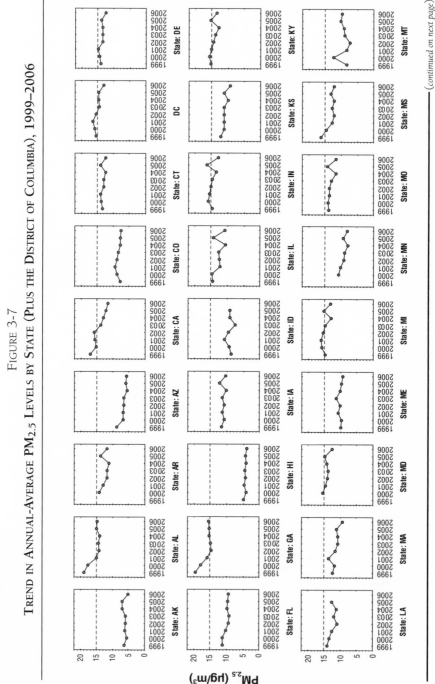

(continued on next page)

(continued from previous page)

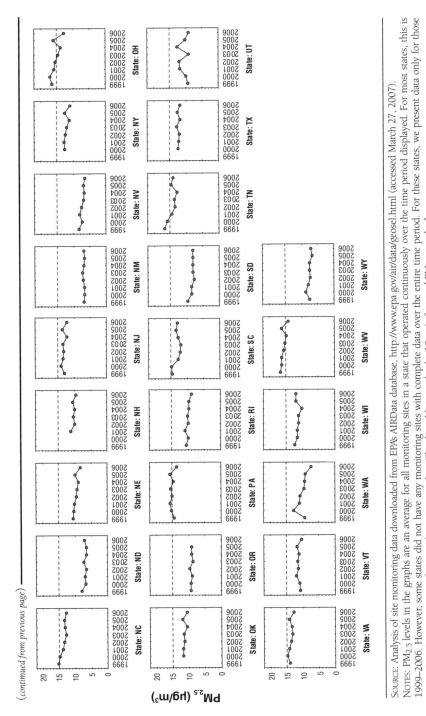

SOURCE: Analysis of site monitoring data downloaded from EPA's AIRData database, http://www.epa.gov/air/data/geosel.html (accessed March 27, 2007).

NOTES: $PM_{2.5}$ levels in the graphs are an average for all monitoring sites in a state that operated continuously over the time period displayed. For most states, this is 1999–2006. However, some states did not have any monitoring sites with complete data over the entire time period. For these states, we present data only for those years when there is a group of sites with continuous data. The dotted line marks the 15 $\mu g/m^3$ annual $PM_{2.5}$ standard.

Appendix to Chapter 6

How We Estimated the Number of People
Living in Areas That Violate Pollution Standards

Although the nation has about a thousand ozone and $PM_{2.5}$ monitors, most people, even in monitored counties, live miles to tens of miles from the nearest monitoring site. Thus, there will always be uncertainties as to the number of people living in areas that violate a pollution standard. Nevertheless, we can use a mathematical technique called interpolation to estimate pollution levels in between monitors. Here's an example: Imagine we found that $PM_{2.5}$ averaged 18 µg/m³ in downtown Los Angeles and 10 µg/m³ in Santa Monica, fifteen miles to the west. By linear interpolation, we would assume that $PM_{2.5}$ averaged 14 µg/m³ halfway between the two sites, and 16 µg/m³ one-quarter of the distance from downtown Los Angeles to Santa Monica.

We use a somewhat more sophisticated, but still straightforward, set of techniques to estimate pollution levels: an octal search followed by an inverse-distance-weighted average. The technique is implemented as follows:

1. Imagine dividing the area around the center of each census tract in the United States into quadrants, and then dividing those quadrants in half again into eight "octants." These would be shaped like wedges whose points all meet at the geographic center of the census tract.

2. In each octant, find the pollution monitor nearest the center of the census tract, or exclude a given octant if there is no monitor within 350 miles. This concludes the octal search.

3. Then average the $PM_{2.5}$ values from each of the monitors, weighting them by the inverse square of the distance from the center of the census tract. This means that, given two monitors, one that is twice as far away will be weighted one-fourth as much. One that is three times as far away will be weighted one-ninth as much, and so on. The weighted average of the nearest monitor in each octant is the estimated $PM_{2.5}$ level for that census tract.

A total of 726 counties, representing 73 percent of the United States' population, had $PM_{2.5}$ monitors with valid data for 2001–3. This poses two problems: First, it means we can't know for sure what $PM_{2.5}$ levels are in areas that are far from a monitor. Second, it makes it difficult to estimate $PM_{2.5}$ at the edges of monitored areas, because it presumably declines to some regional background level at some point.

But if $PM_{2.5}$ happens to be high at the edge of a given monitored area, our method would assume it remains high for hundreds of miles in some cases. We know this is an absurd result, because this generally doesn't happen in regions that have monitors in urban, suburban, and rural areas, where we can compare relative $PM_{2.5}$ levels in these different types of regions. For example, while 18 percent of all $PM_{2.5}$ monitors violated the annual standard in 2003, only 2 percent of monitors in rural areas and 3 percent outside metropolitan statistical areas (MSAs) did.

Given this result, we take the following approach to ensure that we don't understate $PM_{2.5}$ levels in areas far from a monitor: To create a background $PM_{2.5}$ level for each unmonitored county, we create a "pseudo-monitor" with a $PM_{2.5}$ level slightly higher than the levels measured at actual rural monitors in various regions of the country—so, for example, for 2003 we use a background of 13 $\mu g/m^3$ in the Midwest, South, and Northeast, 9 $\mu g/m^3$ in the Southwest and Northwest, and 10 $\mu g/m^3$ in California. These are about 1–2 $\mu g/m^3$ greater than actual rural levels in these areas

If we stop there, we will, of course, conclude that hardly any unmonitored counties have areas that violate the annual $PM_{2.5}$ standard (there are, in fact, a few unmonitored counties that do, based on interpolation from nearby monitored counties with high $PM_{2.5}$ levels). This might be true, because regulators search for high-pollution areas when picking locations for new monitors. But to be on the safe side, we assume that in the unmonitored counties of a given state, the number of people living in areas that violate the annual $PM_{2.5}$ standard is equal to the population in those unmonitored counties, multiplied by the percentage of monitors in non-MSA areas of the state that violate the standard. Because few non-MSA monitors violate the $PM_{2.5}$ or 1-hour ozone standards, this results in a modest increase in our population estimate for $PM_{2.5}$ and 1-hour ozone. It results in a larger increase for 8-hour ozone, as we now discuss.

Estimating the population in areas that violate the 1-hour or 8-hour ozone standards adds an extra challenge: The ozone standard isn't based on annual averages, but on daily ozone levels. For example, the 1-hour standard is based on the fourth-highest daily, 1-hour peak reading over the most recent three years. If that fourth-highest reading is greater than 0.125 ppm, the monitor violates the standard. In practice, this means a monitor violates the standard if it averages more than one day per year exceeding 0.125 ppm during any consecutive three-year period.

This poses a problem for determining what ozone level to use for each monitor. For example, in a given metropolitan area, the fourth-highest 1-hour reading during 2001–3 could have occurred on a different date at each monitoring site. For our analysis, we simply use the fourth-highest 1-hour value for each monitor, regardless of the date on which it occurred. We were not able for this study to run simulations based on actual daily ozone readings. Nevertheless, in those areas with monitors, our method ensures that, if anything, we will overestimate the extent of the geographic area over which the ozone standard is violated, and therefore the population residing in violating areas.

To see that this is so, imagine two ozone monitors—call them X and Y—spaced ten miles apart. Let's say that on its fourth-highest day, X recorded an ozone level of 0.130 ppm, while Y recorded a value of 0.120 ppm. Now assume that Y's fourth-highest day occurred on a different date than X's, and that the ozone values were reversed on that day. Based on the actual fourth-highest readings, no more than half the area between the sites could have violated the 1-hour standard. But we use the higher of the two values for both X and Y, and therefore consider the entire region between the two monitors to violate the 1-hour standard. Table 6-1 summarizes this example.

TABLE 6-1

ILLUSTRATION OF POPULATION ESTIMATION METHOD FOR OZONE

	X	Y	Result
Ozone level on date of X's fourth-highest reading (ppm)	0.130	0.120	Half of area between X and Y violates standard.
Ozone level on date of Y's fourth-highest reading (ppm)	0.120	0.130	Half of area between X and Y violates standard.
Values actually used for population determination	0.130	0.130	Entire area between X and Y violates standard.

For ozone, 722 counties, representing 71 percent of Americans, had at least one monitor with valid data during 2001–3. For unmonitored counties, we use a background value of 0.080 ppm for the 1-hour standard and 0.060 ppm for the 8-hour standard. Just as for $PM_{2.5}$, for each state we estimate the number of people living in areas of unmonitored counties that violate the standard by multiplying the population in these counties by the percentage of non-MSA monitors in a given state that violate either standard.

We performed several checks to make sure the method gives reasonable results for both monitored and unmonitored counties. For example (the results below are as of the end of 2003):

- Some counties have several ozone monitors, but often only a portion of these violate the 8-hour standard. For example, in Los Angeles County 57 percent of monitors violate the standard and these are located in the northern and eastern portions of the county. We estimate that 40 percent of L.A. County's residents live in areas that violated the standard as of 2003. The western portion of the county is the most densely populated and includes all the monitors that attain the 8-hour standard. Thus, even though 43 percent of the monitors attain the standard, you'd expect that more than 43 percent of the county's population would live in areas that attain the standard, and this is exactly what we find with our method.

Maricopa County (Phoenix), Arizona, violates the 8-hour standard at 10 percent of its monitoring locations, and we estimate that 8 percent of residents live in areas that are in violation.

Essex County (Boston), Massachusetts, violates 8-hour standard at 66 percent of its monitoring locations, and we estimate that 65 percent of residents live in areas that are in violation.

Mecklenburg County (Charlotte), North Carolina, violates the 8-hour standard at 66 percent of its monitoring locations, and we estimate that 88 percent of residents live in areas that are in violation.

- The Midwest and Northeast are unlike the West and most of the South in that 8-hour-ozone violations can occur over large areas, sometimes whole states. Indeed, even though not all counties in Connecticut, New Jersey, Delaware, Maryland, or Massachusetts have ozone monitors, we counted all people in these five states as living in areas that violated the 8-hour standard (for 2001–3), because all of their non-MSA monitors were in violation. Table 6-2 shows the violation rates we used for all fifty states plus the District of Columbia. Entries in bold indicate states that either have no monitors in non-MSA counties, or in which all counties are part of an MSA. In these cases, we use the average violation rate for all monitors in the state.

- In the western United States outside of California, hardly any monitoring locations, even in metropolitan areas, violate the 8-hour ozone standard. As expected, our method counts virtually no people in unmonitored counties in the West as living in areas that violate the 8-hour or 1-hour standards.

- More than 99 percent of Californians live in counties that have ozone monitors, and California is, overall, the most heavily monitored state in the nation. This makes sense, given that it has the areas with the highest pollution levels and the most

TABLE 6-2

EIGHT-HOUR OZONE VIOLATION RATES DURING 2001–3 IN
NON-MSA COUNTIES, BY STATE (PLUS THE DISTRICT OF COLUMBIA)

State	Violation Rate	State	Violation Rate
AK	0%	MT	0%
AL	0%	NC	47%
AR	0%	ND	0%
AZ	0%	NE	0%
CA	39%	NH	0%
CO	0%	NJ	100%
CT	100%	NM	0%
DC	100%	NV	0%
DE	100%	NY	60%
FL	0%	OH	67%
GA	25%	OK	0%
HI	0%	OR	0%
IA	0%	PA	83%
ID	0%	RI	100%
IL	0%	SC	13%
IN	83%	SD	0%
KS	0%	TN	40%
KY	7%	TX	0%
LA	14%	UT	0%
MA	100%	VA	33%
MD	100%	VT	0%
ME	71%	WA	0%
MI	75%	WI	20%
MN	0%	WV	0%
MO	0%	WY	0%
MS	0%		

NOTES: For states with no non-MSA counties, we used the average violation rate for all monitors in the state.

frequent exceedances of federal pollution standards. As a result, unmonitored counties do not affect our results for California.

- Because of a unique situation, we made one ad hoc modification to our estimation method for the 8-hour standard. Brooklyn

(Kings County), New York, is by far the most populous urban county without an ozone monitor. With 2.5 million people, it is the seventh-most populous county in the country. Since 60 percent of non-MSA monitors in New York State exceeded the 8-hour standard as of 2003, by our method, we would count 60 percent of the people in Brooklyn as living in areas that are in violation. But the counties adjacent to Brooklyn—Queens and Manhattan—both have ozone monitors, all of which attained the 8-hour standard for 2001–3.[1] Thus, we believe it unlikely that any part of Brooklyn violates the 8-hour standard, and we don't include any residents of Brooklyn in our count of people living in areas that violate the standard.

The results for other populous, but unmonitored, counties seem reasonable, so this is the only ad hoc modification we make. For example, Nassau County, New York (population 1.3 million), lies between two monitored counties: Queens and Suffolk. All monitors in Queens attain the 8-hour standard, while two out of three in Suffolk violate. By our interpolation method, we count 60 percent of Nassau's residents as living in areas that are in violation, which is probably in the right ballpark. Hennepin County, Minnesota (population 1.1 million), is in a state where no monitors violate the standard, as is Honolulu County, Hawaii (population 876,000), so no one in these counties is included in our population count for violating areas. Essex County, New Jersey (population 793,000), is in a state where all monitors violate the standard, so we count all people in Essex as living in an area that is in violation.

The estimation method appears to be relatively robust for counties with pollution monitors. The uncertainties are, of course, greater for unmonitored counties, but this mainly affects 8-hour ozone, since nonattainment is rare and generally more localized for the 1-hour ozone and the $PM_{2.5}$ standards. But within the limitations of the data, we believe our estimates to be reasonable and conservative.

Attainment of ozone and $PM_{2.5}$ standards is based on the most recent three consecutive years of data from a given monitoring site. In order to maximize geographic coverage, we include monitors that had one or two years of valid data as well. This can introduce some error into our estimates, particularly for ozone, because ozone levels can vary substantially from year to year.

We assumed that the population of each county was constant during the years for which we estimate population pollution exposure (we used 2004 population). This could add a small error to our estimated trends, since population has been growing overall and has changed at different rates in different counties. However, using a constant population means that changes in our estimates are solely due to changes in pollution levels, rather than to changes in a county's total population. Since our estimates span only four years, the effect is small, in any case. County population data were downloaded from the Census Bureau.[2]

Notes

Introduction

1. Polling data on Americans' attitudes about air pollution are discussed in chapters 5 and 7.

2. In September 2006 EPA adopted a more stringent standard for 24-hour $PM_{2.5}$ levels. Twenty-four percent of monitoring sites violate this new standard. EPA also proposed a tougher 8-hour ozone standard in June 2007. EPA is currently considering a range for the new ozone standard that would put between 32 and 97 percent of monitoring sites in violation. It will be several years before EPA "designates" nonattainment areas under these new standards, so we base the pollution trend information in this section on currently enforced standards. We discuss pollution levels relative to current and new standards in greater detail in chapters 2 and 3.

3. Standards for $PM_{2.5}$ and 8-hour ozone did not, of course, exist in 1980. But since we have national monitoring data for 1980, we can go back and calculate what the violation rate would have been had current standards existed then.

4. PM was originally measured as "total suspended particulates" (TSP; roughly equivalent to PM_{30}). As health concerns focused on finer particulates, regulators began measuring finer fractions of PM. PM_{10} refers to PM up to 10 microns (10 millionths of a meter) in diameter, while $PM_{2.5}$ is PM up to 2.5 microns in diameter. For comparison, a typical human hair is roughly 70 microns in diameter.

Chapter 1: Air Quality Trends Before and After the Clean Air Act of 1970

1. I. M. Goklany, *Clearing the Air: The Real Story of the War on Air Pollution* (Washington, D.C.: Cato, 1999).

2. Ibid.

3. It is also worth noting that people in Western societies generally did not begin to see air pollution as a problem until the early twentieth century. During the 1800s, factory smokestacks were perceived to represent jobs, progress, and better lives. Most

medical authorities considered smoke to be healthful, as it was believed to drive away disease. See ibid.

4. J. A. Tarr, *The Search for the Ultimate Sink: Urban Pollution in Historical Perspective* (Akron, Ohio: University of Akron Press, 1996).

5. J. H. Ludwig, G. B. Morgan, and T. B. McMullen, "Trends in Urban Air Quality," *EOS* 51 (1970): 468–75.

6. We were not able to locate motor vehicle registrations for L.A. County in 1920. However, those in California as a whole grew by a factor of five from 1920 to 1940; J. E. Krier and E. Ursin, *Pollution and Policy: A Case Essay on California and Federal Experience with Motor Vehicle Air Pollution, 1940–1975* (Los Angeles: University of California, 1977); Pardee Legal Research Center, *California Motor Vehicle Registration and Accident Statistics*, University of San Diego, http://www.sandiego.edu/lrc/vehicle.html#califstatis (accessed July 15, 2006).

7. U.S. Environmental Protection Agency, *National Air Quality and Emission Trends Report, 1976* (Washington, D.C.: Government Printing Office, 1977); U.S. Environmental Protection Agency, *National Air Quality and Emission Trends Report, 1983* (Washington, D.C.: Government Printing Office, 1984).

8. H. Schimmel and T. J. Murawski, "SO_2—Harmful Air Pollutant or Air Quality Indicator?" *Journal of the Air Pollution Control Association* 25, no. 7 (1975): 739–40.

9. To be included in the analysis, a monitoring site had to have at least ten days in every month (from 1963 to 1978) in which there were at least four hourly ozone readings taken between 10 a.m. and 6 p.m. Once these sites were selected, a monitoring day at a given site was included in the analysis only if it met these same criteria. The number of days exceeding 0.1 ppm in a given month was estimated by multiplying the measured number of exceedances by the ratio of the total number of days in a month to the number of days for which there were valid measurements. Thus, if a site had ten exceedances in June out of twenty-five days measured, the estimated exceedances for that month would be 10 x (30/25) = 12. The estimated exceedances by month were then summed to get annual totals. We analyzed the data by month in order to control for the effects of weather on ozone formation (much more ozone is formed in the warmer months than in the cooler months).

10. I. Goklany, "Empirical Evidence Regarding the Role of Nationalization in Improving U.S. Air Quality," in *The Common Law and the Environment: Rethinking the Statutory Basis for Modern Environmental Law*, ed. R. E. Meiners and A. P. Morriss (Lanham, Md.: Rowman & Littlefield, 2000).

11. Ibid.

12. See discussion in chapter 1 of Goklany, *Clearing the Air*.

13. Ibid.

14. See, for example, J. H. Adler, "Fables of the Cuyahoga: Reconstructing a History of Environmental Protection," *Fordham Environmental Law Journal* 14 (2002): 89–146.

15. S. Moore and J. L. Simon, *It's Getting Better All the Time: 100 Greatest Trends of the Last 100 Years* (Washington, D.C.: Cato, 2000).

16. Ibid.

17. S. Peltzman, *Regulation and the Natural Progress of Opulence* (Washington, D.C.: AEI-Brookings Joint Center for Regulatory Studies, 2004), http://www.aei-brookings.org/admin/authorpdfs/page.php?id=1144 (accessed November 27, 2006).

18. The lead content of the fuel supply peaked around 1970, declined slightly from 1970 to 1979, and then declined more than 90 percent between 1979 and 1987. Lead was completely eliminated from gasoline in 1995. L. M. Gibbs, *How Gasoline Has Changed* (Warrendale, Penn.: Society of Automotive Engineers, October 1993); L. M. Gibbs, *How Gasoline Has Changed II—The Impact of Air Pollution Regulations* (Warrendale, Penn.: Society of Automotive Engineers, October 1996).

19. Environmental Protection Agency, *AirTrends*, http://www.epa.gov/airtrends/ (accessed November 15, 2006).

20. For lead levels in gasoline, see Gibbs, *How Gasoline Has Changed*; Gibbs, *How Gasoline Has Changed II*. For ambient lead trends from the 1970s to mid-1990s, see U.S. Environmental Protection Agency, "Air Quality Trends," in *National Air Quality and Emission Trends Report, 1995*, chapter 2, http://www.epa.gov/air/airtrends/aqtrnd95/report/files/chapt2.pdf (accessed August 16, 2006).

21. Although the 8-hour ozone standard wasn't adopted until 1997, because ozone is measured hourly, we were able to use pre-1997 data to determine the long-term trend in 8-hour-average ozone levels.

22. EPA did not require national monitoring of $PM_{2.5}$ until 1999. However, a special study called the Inhalable Particulate Network collected $PM_{2.5}$ data in ninety metropolitan areas from 1979 to 1983.

23. Bureau of Transportation Statistics, *National Transportation Statistics*, U.S. Department of Transportation, http://www.bts.gov/publications/national_transportation_statistics/ (accessed November 15, 2006); http://www.bts.gov/publications/national_transportation_statistics/2005/index.html (accessed August 16, 2006).

24. Bureau of Economic Analysis, *National Economic Accounts*, U. S. Department of Commerce, http://www.bea.doc.gov/bea/dn/nipaweb/SelectTable.asp (accessed November 15, 2006); Energy Information Administration, *Annual Energy Review 2005*, July 2006, http://www.eia.doe.gov/emeu/aer/pdf/aer.pdf (accessed November 15, 2006).

25. For each site, the annual-average pollutant level was constructed by averaging monthly average levels.

26. The number of monitoring sites included in the average ranged from five to seven, depending on the pollutant.

27. CARB estimates that about 75 percent of these compounds come from gasoline vehicles, and more than 80 percent come from mobile sources; California Air Resources Board, *The California Almanac of Emissions and Air Quality, 2005*,

http://www.arb.ca.gov/aqd/almanac/almanac05/almanac2005all.pdf (accessed August 16, 2006).

28. California Department of Transportation, Division of Transportation System Information, *California Motor Vehicle Stock, Travel and Fuel Forecast*, November 2004, http://www.dot.ca.gov/hq/tsip/otfa/mtab/MVSTAFF/MVSTAFF04.pdf (accessed August 16, 2006).

29. Ibid.

30. See, for example, N. Aleksic, G. Boynton, G. Sistla, and J. Perry, "Concentrations and Trends of Benzene in Ambient Air over New York State during 1990–2003," *Atmospheric Environment* 39 (2005): 7894–7905; R. Oomen, J. Jauser, D. Dayton, and G. Brooks, "Evaluating HAP Trends: A Look at Emissions, Concentrations, and Regulation Analyses for Selected Metropolitan Statistical Areas" (paper, Fourteenth International Emission Inventory Conference, U.S. EPA, Las Vegas, Nev., April 12–14, 2005).

Chapter 2: Ozone: Historic Trends and Current Conditions

1. EPA recently tightened the 24-hour $PM_{2.5}$ standard, reducing it from 65 $\mu g/m^3$ to 35 $\mu g/m^3$. This will increase the fraction of monitors that violate $PM_{2.5}$ standards from 14 percent to 24 percent (based on $PM_{2.5}$ data for 2004–6). EPA plans to issue final "designations" under the new standard—that is, the official list of regions in violation of the standard—in 2010, at which time it will become the most widely violated federal air pollution standard. However, EPA also proposed a tougher 8-hour ozone standard in June 2007. Even the least stringent standard EPA is considering would once again make ozone the most widely violated pollution standard once designations under that standard are finalized, probably in 2011 or 2012.

2. U. L. McFarling and M. Bustillo, "2005 Vying with '98 as Record Hot Year—Los Angeles Times," *Los Angeles Times*, December 16, 2005; National Oceanic and Atmospheric Administration, "NOAA Reports 2006 Warmest Year on Record for U.S.," press release, January 9, 2007, http://www.noaanews.noaa.gov/stories2007/s2772.htm (accessed March 15, 2007).

3. U.S. Environmental Protection Agency, "2007 Proposed Revisions to Ground-Level Ozone Standards," http://earth1.epa.gov/air/ozonepollution/naaqsrev2007.html (accessed July 28, 2007).

4. S. Reynolds, C. L. Blanchard, and S. D. Ziman, "Understanding the Effectiveness of Precursor Reductions in Lowering 8-Hr Ozone Concentrations," *Journal of the Air & Waste Management Association* 53 (2003): 195–205; S. Reynolds, C. L. Blanchard, and S. D. Ziman, "Understanding the Effectiveness of Precursor Reductions in Lowering 8-Hour Ozone Concentrations—Part II. The Eastern United States," *Journal of the Air & Waste Management Association* 54 (2004): 1452–70; D. A. Winner and G. R. Cass, "Effect of Emissions Control on the Long-Term Frequency Distribution of Regional Ozone Concentrations," *Environmental Science & Technology* 34 (2000): 2612–17.

5. Winner and Cass, "Effect of Emissions Control on the Long-Term Frequency Distribution of Regional Ozone Concentrations."

6. J. M. Heuss, D. F. Kahlbaum, and G. T. Wolff, "Weekday/Weekend Ozone Differences: What Can We Learn from Them?" *Journal of the Air & Waste Management Association* 53 (2003): 772–88.

7. Based on hourly ozone and NOx monitoring data for 1997–2001 downloaded from the California Air Resources Board's Web site, http://www.arb.ca.gov/aqd/aqdcd/aqdcddld.htm (accessed September 26, 2003).

8. C. L. Blanchard and S. J. Tannenbaum, "Weekday/Weekend Differences in Ambient Air Pollutant Concentrations in Atlanta and the Southeastern United States," *Journal of the Air & Waste Management Association* 56 (2006): 271–84.

9. R. Torres-Jardon and T. C. Keener, "Evaluation of Ozone-Nitrogen Oxides-Volatile Organic Compound Sensitivity of Cincinnati, Ohio," *Journal of the Air & Waste Management Association* 56 (2006): 322–33.

10. E. M. Fujita, W. R. Stockwell, D. E. Campbell, et al., "Evolution of the Magnitude and Spatial Extent of the Weekend Ozone Effect in California's South Coast Air Basin 1981–2000," *Journal of the Air & Waste Management Association* 53 (2003): 864–75; D. R. Lawson, "The Weekend Effect—The Weekly Ambient Emissions Control Experiment," *Environmental Manager*, July 2003, 17–25; L. C. Marr and R. A. Harley, "Modeling the Effect of Weekday-Weekend Differences in Motor Vehicle Emissions on Photochemical Air Pollution in Central California," *Environmental Science & Technology* 36 (2002): 4099–4106; L. C. Marr and R. A. Harley, "Spectral Analysis of Weekday-Weekend Differences in Ambient Ozone, Nitrogen Oxide, and Non-Methane Hydrocarbon Time Series in California," *Atmospheric Environment* 36 (2002): 2327–35; B. K. Pun and C. Seigneur, "Day-of-Week Behavior of Atmospheric Ozone in Three U.S. Cities," *Journal of the Air & Waste Management Association* 53 (2003): 789–801; Reynolds, Blanchard, and Ziman, "Understanding the Effectiveness of Precursor Reductions—Part II"; Torres-Jardon and Keener, "Evaluation of Ozone-Nitrogen Oxides-Volatile Organic Compound Sensitivity of Cincinnati, Ohio"; G. Yarwood, T. E. Stoeckenius, J. G. Heiken, and A. M. Dunker, "Modeling Weekday/Weekend Ozone Differences in the Los Angeles Region for 1997," *Journal of the Air & Waste Management Association* 53 (2003): 864–75.

11. NOx is shorthand for the sum of nitric oxide (NO) and nitrogen dioxide (NO_2). Most NOx is emitted as NO, but is oxidized to NO_2 in the atmosphere. This discussion of ozone formation is summarized from N. Carslaw and D. Carslaw, "The Gas-Phase Chemistry of Urban Atmospheres," *Surveys in Geophysics* 22 (2001): 31–53, and J. H. Seinfeld, "Urban Air Pollution: State of the Science," *Science* 243 (1989): 745–52.

12. Readers with a background in air pollution studies or chemistry will observe that I am using a nonstandard notation. My goal is to make this discussion understandable to lay readers.

13. It seems paradoxical that reducing NOx could speed up reaction (4), because this reaction requires NOx. But recall that in a VOC-limited situation, there's lots of NOx compared to VOCs, so the rate of reaction (4) is not limited by the availability of NOx, but by the availability of peroxy radicals (VOC-OO). However, when there is excess NOx around, fewer peroxides get formed, because NOx competes for the OH radicals that create peroxides from VOCs. Reducing NOx thus increases the rate of peroxide formation, which ultimately increases the rate of ozone formation.

Chapter 3: Particulates: Historic Trends and Current Conditions

1. U.S. Environmental Protection Agency, "Air Trends: Particulate Matter," http://www.epa.gov/airtrends/pm.html (accessed May 3, 2007).

2. This is less than the number of counties EPA has classified as being out of attainment with federal $PM_{2.5}$ standards. There are two reasons for this. First, EPA classifies some counties as being out of attainment not because the counties themselves violate federal standards, but because emissions from them are believed to contribute to nonattainment in nearby areas. Second, some counties that were out of attainment in, say, 2002 or 2003 have since come into attainment, but have not yet been officially reclassified by EPA as being in attainment.

3. U.S. Environmental Protection Agency, *Fact Sheet: Final Revisions to the National Ambient Air Quality Standards for Particle Pollution (Particulate Matter)*, September 21, 2006, http://epa.gov/particles/pdfs/20060921_factsheet.pdf (accessed November 27, 2006).

4. U.S. Environmental Protection Agency, *Inhalable Particulate Network Report: Operation and Data Summary* (Mass Concentrations Only), Volume I, April 1979–December 1982, by D. O. Hinton, J. M. Sune, J. C. Suggs, and W. F. Barnard (Washington, D.C.: Government Printing Office, November 1984); U.S. Environmental Protection Agency, *Inhalable Particulate Network Report: Data Summary (Mass Concentrations Only), Volume III, January 1983–December 1984*, by D. O. Hinton, J. M. Sune, J. C. Suggs, and W. F. Barnard (Washington, D.C.: Government Printing Office, April 1986).

5. We applied one correction to the raw IPN values, because they were measured by a different technique than current $PM_{2.5}$ measurements. When EPA created the $PM_{2.5}$ standards in 1997, it also created a new "federal reference method" (FRM) for measuring $PM_{2.5}$. All else equal, the FRM gives higher $PM_{2.5}$ readings than the dichotomous samplers used to collect the IPN data in the early 1980s. Researchers at the California Air Resources Board (CARB) collected concurrent measurements with both methods and concluded that given the same actual $PM_{2.5}$ levels in air, the FRM gives $PM_{2.5}$ readings about 14 percent higher than the dichotomous samplers used before 1999. The reason for the difference is that some components of $PM_{2.5}$,

particularly nitrates and some organic compounds, are "semivolatile"—that is, they remain solid at cooler temperatures but evaporate at warmer temperatures. The FRM samplers are held at a relatively cool constant temperature and therefore retain the semivolatile components of PM. The older samplers were not cooled, resulting in evaporation of some of the PM after it was collected but before it was analyzed. In order to correct for this bias, we increased IPN measurements from the western half of the United States by 14 percent and by 7 percent in the eastern half. The first figure is the actual magnitude of the bias found by CARB researchers for $PM_{2.5}$ measured by FRM samplers relative to dichotomous samplers sitting at the same location. We applied a smaller correction to the eastern data, because eastern $PM_{2.5}$ includes fewer nitrates and more sulfates than western $PM_{2.5}$. Sulfates do not evaporate, and we would therefore expect eastern IPN $PM_{2.5}$ measurements to be less biased (relative to the FRM) than western IPN $PM_{2.5}$ measurements. However, recent measurements in Ohio have shown that $PM_{2.5}$ in that region might also contain a substantial fraction of semivolatile species. These data compared FRM samplers with a third type of PM sampler—a tapered element oscillating microbalance, or TEOM. Thus, it is possible that our 7 percent correction for eastern $PM_{2.5}$ is too small. If so, we would underestimate the percentage reduction in $PM_{2.5}$ since the early 1980s. D. P. Connell, J. A. Withum, S. E. Winter, et al., "The Steubenville Comprehensive Air Monitoring Program (SCAMP): Analysis of Short-Term and Episodic Variations in PM2.5 Concentrations Using Hourly Air Monitoring Data," *Journal of the Air & Waste Management Association* 55 (2005): 559–73; N. Motallebi, J. Taylor, A. Clinton, et al., "Particulate Matter in California: Part 1—Intercomparison of Several PM2.5, PM10–2.5, and PM10 Monitoring Networks," *Journal of the Air & Waste Management Association* 53 (2003): 1509–16.

6. The exceedance rate for the early 1980s is based on the eighty-seven IPN sites. The exceedance trend for 1999–2006 is based on the 291 continuous sites. As discussed in the text, the eighty-seven IPN sites appear to provide a representative national average, so it seems reasonable to use them for an estimate of the national $PM_{2.5}$ exceedance rate back in the early 1980s.

7. Note that in order to violate the annual $PM_{2.5}$ standard, a site must average more than 15 µg/m^3 over a consecutive three-year period, so the trend in the violation rate begins in 2001, rather than 1999.

8. W. J. Parkhurst, R. L. Tanner, F. P. Weatherford, et al., "Historic PM2.5/PM10 Concentrations in the Southeastern United States—Potential Implications of the Revised Particulate Matter Standard," *Journal of the Air & Waste Management Association* 49 (1999): 1060–67.

9. EPA did not correct the pre-1999 $PM_{2.5}$ data for bias between the old and new measurement methods. We have multiplied the pre-1999 results by 1.07 to correct for this bias; U.S. Environmental Protection Agency, *The Particle Pollution Report*, December 2004, http://www.epa.gov/air/airtrends/aqtrnd04/pm.html (accessed November 27, 2006).

10. PM$_{2.5}$ data from California Air Resources Board, *2006 Air Quality Data CD*. Data were collected by dichotomous sampler from 1988 to 1998, and by the Federal Reference Method thereafter. Dichotomous sampler results were adjusted upward by 13.6 percent, based on dichot-FRM comparisons by CARB staff. See Motallebi, Taylor, Croes, et al., "Particulate Matter in California: Part 1."

11. As noted earlier, EPA recently adopted a tougher standard of 35 μg/m^3. However, EPA won't begin enforcing this standard until 2010.

12. However, most areas measure PM$_{2.5}$ only every third or sixth day. In these cases, EPA considers the ninety-eighth percentile reading to be, respectively, the highest and second-highest reading each year.

13. U.S. Environmental Protection Agency, *Latest Findings on National Air Quality: 2002 Status and Trends*, August 2003, http://www.epa.gov/air/airtrends/aqtrnd02/2002_airtrends_final.pdf (accessed August 23, 2006).

14. U.S. Environmental Protection Agency, *Fact Sheet: Final Revisions to the National Ambient Air Quality Standards for Particle Pollution*.

15. E. Malek, T. Davis, R. S. Martin, and P. J. Silva, "Meteorological and environmental aspects of one of the worst national air pollution episodes (January, 2004) in Logan, Cache Valley, Utah, USA," *Atmospheric Environment* 79 (2006): 108–22.

Chapter 4: Why Air Pollution Will Continue to Decline

1. U.S. Environmental Protection Agency, *Air Emissions Trends—Continued Progress through 2004*, http://www.epa.gov/airtrends/2005/econ-emissions.html (accessed August 23, 2006).

2. U.S. Environmental Protection Agency, "1970–2002 Average Annual Emissions, All Criteria Pollutants," http://www.epa.gov/ttn/chief/trends/index.html#tables (accessed August 23, 2006).

3. Ibid.

4. D. D. Parrish, "Critical Evaluation of US On-Road Vehicle Emission Inventories," *Atmospheric Environment* 40 (2006): 2288–2300.

5. E. M. Fujita, D. E. Campbell, B. Zielinska, et al., "Diurnal and Weekday Variations in the Source Contributions of Ozone Precursors in California's South Coast Air Basin," *Journal of the Air & Waste Management Association* 53 (2003): 844–63; J. G. Watson, J. C. Chow, and E. M. Fujita, "Review of Volatile Organic Compound Source Apportionment by Chemical Mass Balance," *Atmospheric Environment* 32 (2001): 1567–84.

6. U.S. Environmental Protection Agency, *1999 National Emission Inventory Documentation and Data—Final Version 3.0*, February 2001, http://www.epa.gov/ttn/chief/net/1999inventory.html (accessed September 20, 2006); California Air Resources Board, *CEFS Emissions by Summary Category*, http://www.arb.ca.gov/app/emsinv/ccos/fcemssumcat_cc214.php (accessed August 27, 2006).

7. South Coast Air Quality Management District, *2003 Air Quality Management Plan, Appendix III: Base and Future Year Emission Inventories* (Diamond Bar, Calif.: South Coast Air Quality Management District, August 2003).

8. There are dozens of different types of off-road diesel vehicles and engines, so there is no way to make a simple chart depicting all of the off-road diesel standards. For more information on off-road diesel standards, see U.S. Environmental Protection Agency, *Final Regulatory Impact Analysis: Control of Emissions from Nonroad Diesel Engines*, May 2004, http://www.epa.gov/nonroad-diesel/2004fr/420r04007a.pdf (accessed August 27, 2006).

9. GVWR is the maximum design weight of a vehicle when fully loaded with passengers and cargo.

10. J. G. Calvert, J. B. Heywood, R. F. Sawyer, and J. H. Seinfeld, "Achieving Acceptable Air Quality: Some Reflections on Controlling Vehicle Emissions," *Science* 261 (1993): 37–45; U.S. Environmental Protection Agency, "Control of Air Pollution from New Motor Vehicles: Tier 2 Motor Vehicle Emission Standards and Gasoline Sulfur Control Requirements; Final Rule," *Federal Register* 65, no. 28 (February 10, 2000): 6698–6870, http://frwebgate2.access.gpo.gov/cgibin/waisgate.cgi?WAIS docID=246027345550+8+0+0&WAISaction=retrieve (accessed December 4, 2006); Environmental Protection Agency, *Federal and California Exhaust and Evaporative Emission Standards for Light-Duty Vehicles and Light-Duty Trucks*, February 2000, http://www.epa.gov/otaq/cert/veh-cert/b00001.pdf (accessed August 27, 2006).

11. J. Schwartz, *No Way Back: Why Air Pollution Will Continue to Decline* (Washington, D.C.: American Enterprise Institute, 2003), http://www.aei.org/docLib/2003 0804_4.pdf (accessed August 27, 2006).

12. Based on estimates of miles driven per year from Texas Transportation Institute, *2005 Urban Mobility Study*, "Congestion Data for Your City," http://mobility.tamu.edu/ ums/congestion_data/tables/complete_data.xls.(accessed November 27, 2006).

13. A. J. Kean, R. F. Sawyer, R. A. Harley and G. R. Kendall, *Trends in Exhaust Emissions from In-Use California Light-Duty Vehicles, 1994–2001* (Warrendale, Penn.: Society of Automotive Engineers, 2002).

14. See this as follows: Assume the average car's VOC emissions were equal to 1.0 (in arbitrary units) in 1994. By 2001, these emissions had declined 67 percent to 0.33 (1.00 − 0.67 = 0.33). But gasoline use increased 13 percent, so the decline in total emissions was 1.00 − (0.33 x 1.13) = 0.63, or 63 percent.

15. The California on-road data extend through the 2001 model year. The Phoenix data extend through the 1999 model year. California roadside data were provided by the California Bureau of Automotive Repair. Phoenix data were provided by Tom Wenzel of Lawrence Berkeley National Laboratory.

16. California Air Resources Board, *The California Low-Emission Vehicle Regulations*, January 2006, http://www.arb.ca.gov/msprog/levprog/cleandoc/cleancompletelev-ghgregs11-7.pdf (accessed August 27, 2006); U.S. Environmental Protection Agency,

Regulatory Impact Analysis: Tier 2/Gasoline Sulfur Final Rulemaking (Washington, D.C.: Government Printing Office, December 1999).

17. Traci Watson, "Smoggy Skies Persist Despite Decade of Work," *USA Today*, October 16, 2003, 1.

18. Sierra Club, *Clearing the Air with Transit Spending* (Washington, D.C.: Sierra Club, November 2001).

19. D. Goldberg, "If Smog Isn't Routed, It Returns," *Atlanta Journal-Constitution*, September 1, 2003, 13A.

20. A. W. Gertler, M. Abu-Allaban, W. Coulombe, et al., "Measurements of Mobile Source Particulate Emissions in a Highway Tunnel," *International Journal of Vehicle Design* 27 (2002): 86–93; T. Kirchstetter, A. Strawa, G. Hallar, et al., "Characterization of Particle and Gas Phase Pollutant Emissions from Heavy- and Light-Duty Vehicles in a California Roadway Tunnel" (paper, American Geophysical Union Fall Meeting, San Francisco, December 13–17, 2004).

21. Kirchstetter, Strawa, Hallar, et al., "Characterization of Particle and Gas Phase Pollutant Emissions from Heavy- and Light-Duty Vehicles in a California Roadway Tunnel."

22. A. W. Gertler, J. A. Gillies, W. R. Pierson, et al., *Emissions from Diesel and Gasoline Engines Measured in Highway Tunnels* (Boston: Health Effects Institute, January 2002).

23. R. A. Harley, L. C. Marr, J. K. Lehner, and S. N. Giddings, "Changes in Motor Vehicle Emissions on Diurnal to Decadal Time Scales and Effects on Atmospheric Composition," *Environmental Science and Technology* 39 (2005): 5356–62; J. Yanowitz, R. L. McCormick, and M. S. Graboski, "In-Use Emissions from Heavy-Duty Diesel Vehicles," *Environmental Science & Technology* 34 (2000): 729–40.

24. U.S. Environmental Protection Agency, *Health Assessment Document for Diesel Engine Exhaust* (Washington, D.C.: Government Printing Office, May 2002).

25. U.S. Environmental Protection Agency, "DOJ, EPA Announce One Billion Dollar Settlement with Diesel Engine Industry for Clean Air Violations," press release, October 22, 1998, http://yosemite1.epa.gov/opa/admpress.nsf/b1ab9f485b09897 2852562e7004dc686/93e9e651adeed6b7852566a60069ad2e?OpenDocument (accessed December 4, 2006); Bruce Yandle and Andrew P. Morriss, "Regulation by Litigation: Diesel Engine Emission Control," *Independent Review* 8 (2004): 401–18, http://www.independent.org/pdf/tir/tir_08_3_yandle.pdf.

26. CTC & Associates, *Achieving Compliance with the Diesel Reflash Program*, prepared for the Wisconsin Department of Transportation, August 30, 2006, http://www.dot.wisconsin.gov/library/research/docs/tsrs/tsrdieselreflash.pdf (accessed November 27, 2006).

27. U.S. Environmental Protection Agency, *Health Assessment Document for Diesel Engine Exhaust*.

28. G. Ban-Weiss and R. A. Harley, "On-Road Measurements of Light-Duty Gasoline and Heavy-Duty Diesel Vehicle Emissions Trends" (paper, Sixteenth Annual CRC On-road Emissions Workshop, San Diego, Coordinating Research Council, March 2006)."

29. U.S. Environmental Protection Agency, "1970–2002 Average Annual Emissions, All Criteria Pollutants."

30. Ibid.

31. Ibid.

32. Ibid.

33. U.S. Environmental Protection Agency, *Acid Rain Program 2004 Progress Report*, October 2005, http://www.epa.gov/airmarkets/cmprpt/arp04/2004report.pdf (accessed August 27, 2006).

34. The SIP Call focused on eastern states, because coal is a common electricity fuel there. Power plants are a smaller contributor to NOx in the West, where natural gas and hydroelectric power are more common; U.S. Environmental Protection Agency, *Addendum to the Regulatory Impact Analysis for the NOx SIP Call, FIP, and Section 126 Petitions* (Washington, D.C.: Environmental Protection Agency, September 1998).

35. Environmental Integrity Project and Public Citizen, *America's Dirtiest Power Plants: Plugged into the Bush Administration*, May 2004, http://www.environmentalintegrity.org/pubs/AmericasDirtiest.pdf (accessed August 27, 2006); Public Citizen, "As Rules Are Relaxed, Sulfur Dioxide and Carbon Dioxide Pollution Go Up," press release, May 5, 2004, http://www.citizen.org/pressroom/release.cfm?ID=1706 (accessed August 27, 2006).

36. U.S. Environmental Protection Agency, *Taking Toxics Out of the Air*, August 2000, http://www.epa.gov/oar/oaqps/takingtoxics/airtox.pdf (accessed August 27, 2006); U.S. Environmental Protection Agency, *Rule and Implementation Information for Hazardous Organic NESHAP*, last updated June 27, 2006, http://www.epa.gov/ttn/atw/hon/honpg.html (accessed November 27, 2006).

37. U.S. Environmental Protection Agency, *Taking Toxics Out of the Air*; U.S. Environmental Protection Agency, *Rule and Implementation Information for Petroleum Refineries*, last updated August 25, 2006, http://www.epa.gov/ttn/atw/petrefine/petrefpg.html (accessed November 27, 2006); U.S. Environmental Protection Agency, *Petroleum Refineries: Catalytic Cracking, Catalytic Reforming and Sulfur Plant Units*, last updated August 25, 2006, http://www.epa.gov/ttn/atw/petuuu/petuuupg.html (accessed November 27, 2006).

38. All the specific rules can be downloaded from http://www.epa.gov/ttn/atw/mactfnlalph.html (accessed November 27, 2006).

39. U.S. Environmental Protection Agency, "National Volatile Organic Compound Emission Standards for Architectural Coatings: Final Rule," *Federal Register* 63, no. 176 (September 11, 1998): 48848–87, http://frwebgate4.access.gpo.gov/cgi-bin/waisgate.cgi?WAISdocID=24865012879+0+0+0&WAISaction=retrieve (accessed December 4, 2006); and U.S. Environmental Protection Agency, "National Volatile Organic Compound Emission Standards for Consumer Products: Final Rule," *Federal Register* 63, no. 176 (September 11, 2006): 48819–47, http://www.epa.gov/ttn/atw/183e/cp/fr1193.pdf (accessed December 4, 2006).

40. As will be discussed in chapter 7, even these low risk levels are likely an over-estimate. See EPA's bar charts of estimated risks from HAPs in U.S. Environmental Protection Agency, "Key Risk Assumptions and Limitations," *Technology Transfer Network National Air Toxics Assessment*, August 23, 2006, http://www.epa.gov/ttn/atw/nata/chartrisk.html (accessed September 20, 2006).

41. Information on EPA's NESHAPs for coke oven and hexavalent chromium emissions can be found at U.S. Environmental Protection Agency, *Rule and Implementation Information for Coke Oven Batteries*, March 9, 2006, http://www.epa.gov/ttn/atw/coke/cokepg.html (accessed December 4, 2006); U.S. Environmental Protection Agency, *Rule and Implementation Information for Chromium Electroplating*, November 1, 2006, http://www.epa.gov/ttn/atw/chrome/chromepg.html (accessed December 4, 2006); and U.S. Environmental Protection Agency, *Rule and Implementation Information for Industrial Cooling Towers*, April 19, 2006, http://www.epa.gov/ttn/atw/cool/cooltpg.html (accessed December 4, 2006).

42. We take up the health effects of mercury in chapter 7.

43. U.S. Environmental Protection Agency, *2002 National Emission Inventory Documentation and Data*, http://www.epa.gov/ttn/chief/net/2002inventory.html (accessed December 18, 2006).

44. U.S. Environmental Protection Agency, *Mercury Study Report to Congress Volume II: An Inventory of Anthropogenic Mercury Emissions in the United States*, December 1997, http://www.epa.gov/ttn/oarpg/t3/reports/volume2.pdf (accessed August 27, 2006).

45. U.S. Department of the Interior, Bureau of Mines, *The Materials Flow of Mercury in the United States*, by S. M. Jasinski, 1994, http://pubs.usgs.gov/usbmic/ic-9412/mercury.pdf (accessed August 28, 2006); U.S. Geological Survey, *The Materials Flow of Mercury in the Economies of the United States and the World*, by J. L. Sznopek and T. G. Goonan, 2000, http://pubs.usgs.gov/circ/2000/c1197/c1197.pdf (accessed August 28, 2006).

46. For this estimate, we assumed that two-thirds of the mining emissions were air emissions.

47. U.S. Geological Survey, *The Materials Flow of Mercury in the Economies of the United States and the World*.

48. U.S. Department of the Interior, Bureau of Mines, *The Materials Flow of Mercury in the United States*.

49. J. D. Husar and R. B. Husar, "Trend of Anthropogenic Mercury Flow in Florida, 1930–2000" (paper, Mercury in the Environment: Assessing and Managing Multimedia Risks Meeting, American Chemical Society, Division of Environmental Chemistry, Orlando, Fla., April 7–11, 2002).

50. Based on analysis of monitoring data from continuously operated sites, down-loaded from EPA's Air Quality System (AQS) database, http://www.epa.gov/ttn/airs/airsaqs/detaildata/downloadaqsdata.htm and http://www.epa.gov/ttn/airs/airsaqs/archived%20data/downloadaqsdata-o.htm (accessed January 21, 2006).

51. Ibid.

52. According to EPA estimates for 2002, 91 percent of CO emissions come from motor vehicles and 87 percent come from gasoline vehicles in particular; U.S. Environmental Protection Agency, "1970–2002 Average Annual Emissions, All Criteria Pollutants."

53. See, for example, U.S. Environmental Protection Agency, *1999 National Emissions Inventory.* Also see the California Air Resources Board's emissions inventory estimates for California: California Air Resources Board, "Forecasted Emissions by Summary Category CCOS Domain Planning Projections - v2.14 RF#956PEI," http://www.arb.ca.gov/app/emsinv/ccos/fcemssumcat_cc214.php (accessed March 8, 2006).

54. U.S. Environmental Protection Agency, *Clean Vehicles + Clean Fuel = Cleaner Air,* 2004, http://www.epa.gov/tier2/420f04002.pdf (accessed August 28, 2006); Natural Resources Defense Council, "EPA Touts New Cleaner Cars," *Bush Record,* January 26, 2004, http://www.nrdc.org/bushrecord/2004_01.asp (accessed August 28, 2006).

55. This is based on MOBILE6 runs provided by Dennis Kahlbaum of Air Improvement Resource, using default assumptions for a fleet in an area with a vehicle inspection program. The prediction includes only tailpipe emissions (that is, it excludes non-tailpipe VOC emissions) and includes a weighted average of cold-start and hot-stabilized emissions. Based on a recent validation study of MOBILE6 against various types of real-world data, it appears that MOBILE6 makes relatively accurate predictions for NOx emissions and might overestimate VOC by up to a few tenths of a gram per mile. The validation study included only on-road data from warmed-up automobiles and used on-road data collected from the early 1990s through 2000 for comparison. The accuracy of the model for cold-start emissions—that is, emissions that occur in the first minute or so after a car is started and the catalytic converter has not yet warmed up—is not known. A. Pollack, C. Lindhjem, T. E. Stoeckenius, et al., *Evaluation of the U.S. EPA MOBILE6 Highway Vehicle Emission Factor Model,* Coordinating Research Council, Environ International, March 2004, http://www.crcao.com/reports/recentstudies2004/CRC_E-64_Final_032004.pdf (accessed August 28, 2006).

56. EPA requires an average Tier 2 automobile to emit no more than 0.125 gram/mile VOC+NOx up to 50,000 miles, and no more than 0.16 gram/mile VOC+NOx between 50,000 and 120,000 miles. To estimate emissions of a Tier 2 fleet, we assumed cars emit 0.125 gram/mile from zero through four years of age; emissions then increase linearly up to 0.16 gram/mile at 120,000 miles, or eleven years of age. Emissions are then assumed to rise 20 percent per year up to 150,000 miles or fourteen years of age, which is the assumed life of the vehicle. For mileage versus vehicle age, we used the MOBILE6 inputs, which are similar to the results of the 1995 National Personal Transportation Survey. The results aren't very sensitive to changes in the assumptions. For example, if we assume emissions immediately jump up to 0.16 gram/mile after 50,000 miles and then increase 30 percent per year after 120,000 miles, fleet-average emissions rise from 0.164 gram/mile to 0.185 gram/mile. Or, to put it conversely, emissions would drop by 1.64 grams/mile in the

"base" case, or 1.61 grams/mile in the "high deterioration" case. If we add the further assumption that the average vehicle lasts for 175,000 miles (about seventeen years) instead of 150,000 miles, then, once again assuming 30 percent/year emissions deterioration, fleet-average VOC+NOx emissions would be 0.23 gram/mile, or an 87 percent decrease from the 2005 fleet-average estimate of 1.8 grams/mile. As a further sensitivity test, assume we've overstated the emissions of the current average automobile. Assume instead that the average automobile's emissions were 20 percent lower, or 1.44 grams/mile in 2005. Then, even in the "high-deterioration/high-mileage" case above, we would still expect per-mile VOC+NOx emissions to decline 84 percent with the Tier 2 standards; U.S. Environmental Protection Agency, "Control of Air Pollution from New Motor Vehicles: Tier 2 Motor Vehicle Emission Standards and Gasoline Sulfur Control Requirements; Final Rule"; Bureau of Transportation Statistics, *Highlights of the 2001 National Household Travel Survey* (Washington, D.C.: Department of Transportation, 2003), http://www.bts.gov/publications/highlights_of_the_2001_national_household_travel_survey/pdf/entire.pdf (accessed November 27, 2006).

57. Calculate this as follows: Set both total automobile emissions and gram/mile emissions of the average automobile to an arbitrary value of 1.0 in 2005. Assume average per-mile emissions decline 90 percent over the next twenty years down to a value of 0.1 in 2025. If total driving stays the same, then total emissions will also be 0.1, or a 90 percent reduction. But if total miles of driving increases 80 percent, then total emissions will be 0.18 in 2024 (0.1 x 1.8 = 0.18), or an 82 percent reduction. Thus, even after including substantial growth in driving, total emissions decline 82 percent.

58. U.S. Environmental Protection Agency, *Regulatory Impact Analysis: Heavy-Duty Engine and Vehicle Standards and Highway Diesel Fuel Sulfur Control Requirements*, December 2000, http://www.epa.gov/otaq/highway-diesel/index.htm (accessed August 28, 2006); Schwartz, *No Way Back.*

59. NOx trend is based on running MOBILE6 with default assumptions. MOBILE6 diesel PM trend is from J. Grannell, C. Ho, T. Tang, and M. Claggett, *Analysis of MOBILE6.2's PM Emission Factor Estimating Function* (Washington, D.C.: U.S. Department of Transportation, Federal Highway Administration, 2005), http://www.epa.gov/ttn/chief/conference/ei13/mobile/granell.pdf (accessed November 27, 2006).

60. U.S. Environmental Protection Agency, *Final Regulatory Impact Analysis: Control of Emissions from Nonroad Diesel Engines.*

61. Ibid.

62. U.S. Environmental Protection Agency, *Regulatory Announcement: Environmental Benefits of Emission Standards for Locomotives*, December 1999, http://www.epa.gov/otaq/regs/nonroad/locomotv/frm/42097049.pdf (accessed August 30, 2006); U.S. Environmental Protection Agency, *Regulatory Announcement: Emission Standards for New Nonroad Engines*, September 2002, http://www.epa.gov/otaq/regs/nonroad/2002/f02037.pdf (accessed August 30, 2006); U.S. Environmental Protection

Agency, *Regulatory Announcement: Frequently Asked Questions from Facility Managers and Other Owners of Industrial Spark-Ignition Engines*, September 2002, http://www.epa.gov/otaq/regs/nonroad/2002/f02041.pdf (accessed August 30, 2006).

63. Transportation conformity is a Clean Air Act provision requiring metropolitan areas to demonstrate that planned transportation projects, such as expanding freeway capacity, will not cause an increase in emissions above the level budgeted in the region's CAA compliance plan.

64. P. McClintock, "MOBILE6 vs. On-Road Exhaust Emissions and MOBILE6 Evaporative Credits vs. I/M Gas Cap Failures" (paper, Thirteenth Annual Mobile Sources Clean Air Conference, National Center for Vehicle Emissions Control and Safety, Steamboat Springs, Colo., September 2003); P. McClintock, "Comparing Remote Sensing Emissions Measurements in St. Louis to Emissions Estimates from the MOBILE6 Arterial Roadway Type" (paper, Sixteenth Annual CRC On-Road Emissions Workshop, Coordinating Research Council, San Diego, March 2006).

65. MOBILE6 assumes an improvement of about 8 percent per year, while the on-road and inspection data suggest about a 12 percent improvement each year. The MOBILE6 trend is based on MOBILE6 output provided by Dennis Kahlbaum of Air Improvement Resource, using national-default inputs. For details on real-world automobile emissions improvements, see S. S. Pokharel, G. A. Bishop, D. H. Stedman, and R. Slott, "Emissions Reductions as a Result of Automobile Improvement," *Environmental Science and Technology* 37 (2003): 5097–5101; Schwartz, *No Way Back*.

66. See note 56 for more on the evidence for this.

67. U.S. Environmental Protection Agency, "Projected Annual SO_2 and NOx Emissions, and Projected Costs of CAIR," *Clean Air Interstate Rule*, 2005, http://www.epa.gov/cair/charts_files/cair_emissions_costs.pdf (accessed August 30, 2006).

68. U.S. Environmental Protection Agency, *Fact Sheet—EPA's Clean Air Mercury Rule*, March 15, 2005, http://www.epa.gov/air/mercuryrule/factsheetfin.htm (accessed August 30, 2006).

69. For a list of rules with compliance dates from 2005 onward, see U.S. Environmental Protection Agency, *Emission Standards for Hazardous Air Pollutants*, September 19, 2006, http://www.epa.gov/ttn/atw/mactfnlalph.html (accessed September 20, 2006).

Chapter 5: Exaggerating Air Pollution Levels; Obscuring Positive Trends

1. Frank O'Donnell, "Smog Problems Nearly Double in 2005," Clean Air Watch, press release, November 10, 2005, http://cleanairwatchpressroom.blogspot.com/2005/11/smog-problems-nearly-double-in-2005.html (accessed August 30, 2006).

2. Pennsylvania Department of Environmental Protection, "Number of Ozone Action Days Up from Last Year," press release, September 28, 2005, http://www.ahs.

dep.state.pa.us/newsreleases/default.asp?ID=3643&varQueryType=Detail (accessed August 30, 2006).

3. U.S. Environmental Protection Agency, Region 1, "New England Experienced More Smog Days During Recent Summer," press release, September 26, 2005, http://www.epa.gov/region1/pr/2005/sep/dd050917.html (accessed August 30, 2006).

4. J. Holtz, "A Hot Summer Meant More Smog," *New York Times*, October 2, 2005.

5. U. L. McFarling and M. Bustillo, "2005 Vying with '98 as Record Hot Year," *Los Angeles Times*, December 16, 2005.

6. For example, a survey commissioned by the American Lung Association found that 90 percent of people trust environmental information provided by the ALA, 59 percent of them a "great deal," while 79 percent trust EPA. A 2002 poll commissioned by the Sierra Club found that a majority of Americans trust environmental groups for information on environmental issues. American Lung Association, "Survey Shows Public Trusts EPA to Set Air Pollution Standards," press release, June 16, 1999; Mellman Group, "Memorandum: National Survey Results," prepared for the Sierra Club, June 19, 2002. http://www.wsn.org/issues/sierrapoll.pdf (accessed November 27, 2006).

7. Foundation for Clean Air Progress, *January 2002 National Quorum Results*, Wirthlin Worldwide, January 14, 2002, www.cleanairprogress.org/news/quorum_res_01_14_02.asp (accessed September 1, 2006); Foundation for Clean Air Progress, *Clean Air National Survey Results*, Wirthlin Worldwide, August 2004, http://www.cleanairprogress.org/research/clean_secret_survey.asp (accessed September 1, 2006); Foundation for Clean Air Progress, *Survey of Air Pollution Perceptions: Final Report*, ICR, September 1999, http://www.cleanairprogress.org/research/Perceptions.pdf (accessed September 1, 2006).

8. J. Rauch, "America Celebrates Earth Day 1970—for the 31st Time," *National Journal*, April 29, 2000.

9. M. Baldassare, *PPIC Statewide Survey: Special Survey on Californians and the Environment* (San Francisco: Public Policy Institute of California, July 2004).

10. PollingReport.com, *Environment [Compilation of Survey Results from Nationwide Surveys on Environmental Issues]*, 2002, www.pollingreport.com/enviro.htm (accessed October 28, 2002).

11. Washingtonpost.com, *On Politics: Poll Vault*, http://www.washingtonpost.com/wp-srv/politics/polls/vault/stories/worries110799.htm (accessed November 27, 2006).

12. *State of the Air* is always based on air pollution levels during the three-year period from four to two years before the year in which the report is released. Thus, *State of the Air: 2003* was based on pollution levels during 1999–2001. American Lung Association, *State of the Air: 2003*, http://lungaction.org/reports/stateofthe air2003.html (accessed September 29, 2006).

13. American Lung Association, *State of the Air: 2006*, http://lungaction.org/reports/stateoftheair2006.html (accessed September 29, 2006).

14. U.S. Environmental Protection Agency, "U.S. EPA Downgrades San Joaquin Valley Air," press release, October 23, 2001, http://yosemite.epa.gov/opa/admpress.nsf/bf92f4e7d755207d8525701c005e38d7/ee00e5fd6dca7bbd852570d8005e145a!OpenDocument (accessed September 1, 2006).

15. Ibid.

16. U.S. Public Interest Research Group, *Danger in the Air* (Washington, D.C.: U.S. PIRG, August 2002).

17. Ibid.

18. O'Donnell, "Smog Problems Nearly Double in 2005."

19. J. M. Broder, "Cleaner Air in Los Angeles? Don't Hold Your Breath," *New York Times*, November 14, 2004.

20. All of the pollution results by zip code from ALA's Web site were downloaded on December 15, 2005. Although ALA has since released the 2006 and 2007 editions of its report, as of this writing ALA's 2005 results by zip code can still be downloaded at http://lungaction.org/reports/stateoftheair2005.html.

21. The monitor is located at 6400 Bissonnet Street in Houston.

22. The monitor is located at 2311 Texas Avenue in Houston.

23. American Lung Association, "Air Quality," http://www.lungusa.org/site/pp.asp?c=dvLUK9O0E&b=33691 (accessed September 20, 2006). The 152 million claim comes from American Lung Association, *State of the Air: 2005*, April 2005, http://lungaction.org/reports/stateoftheair2005.html (accessed September 29, 2006).

24. In December 2005, EPA proposed a tougher 24-hour $PM_{2.5}$ standard of 35 $\mu g/m^3$. But this proposal came more than a year and a half after the pollution-exaggeration issue discussed here.

25. ALA's claim comes from American Lung Association, *State of the Air: 2004*, http://lungaction.org/reports/stateoftheair2004.html (accessed September 29, 2006). Actual $PM_{2.5}$ levels come from analysis of site monitoring data downloaded from EPA's AIRData database, http://www.epa.gov/air/data/geosel.html (accessed September 29, 2006).

26. U.S. Environmental Protection Agency, *Air Quality Index Reporting; Final Rule*, August 4, 1999, http://www.epa.gov/ttncaaa1/t1/fr_notices/airqual.pdf (accessed September 4, 2006).

27. Based on all sites with three years of data from 2001 to 2003.

28. U.S. Environmental Protection Agency, *PM Standards Revision—2006*, http://www.epa.gov/air/particlepollution/naaqsrev2006.html (accessed November 21, 2006).

29. The other two are Orange County in California, where one out of four monitors violates the 8-hour ozone standard, and Harris County (Houston), Texas, where all but one monitoring location violate the 8-hour standard.

30. The population data are from the Census Bureau. See chapter 6 for more information on how we estimated the number of people living in locations that violate EPA's pollution standards.

31. J. Heilprin, "EPA Designates 474 Counties as Failing Federal Air Quality Standards," Associated Press, April 15, 2004, http://www.wvbt.com/global/story .asp?s=1789017&ClientType=Printable (accessed September 4, 2006).

32. Six locations had data for all three years. An additional two locations had data for one or two years.

33. R. C. Herguth and C. Sadovi, "Jump at Pump Fueling a Hot Debate," *Chicago Sun-Times*, June 18, 2000, 8.

34. M. Schnurman, "Is a New Stadium Worth the Price?," *Fort Worth Star Telegram*, May 2, 2004, 1F.

35. S. Kiehl, "Cleaner-Fuel Buses Sought to Reduce Air Pollution," *Baltimore Sun*, December 12, 2002, 6B.

36. *Westchester (New York) Journal News*, "Region's Air Quality Improvement Plan Found Lacking," December 3, 1999, 5B.

37. S. Shelton, "Clean Air: Court Rejects Georgia Anti-Smog Proposal," *Atlanta Journal-Constitution*, August 23, 2002.

38. L. Layton, "Clashing Color Alerts Blur the Message," *Washington Post*, May 11, 2003, C04; M. Morrison, "Arizona's Alternative Fuel Incentives Backfire; After $200 Million Mistake, State Reneges on Promise of Tax Credits to Buyers of Altered Vehicles," *Washington Post*, December 11, 2000, A03.

39. A. Nussbaum, "Clean Cars Get Airing of Views; Environment Supporters vs. Auto Industry," *Bergen (New Jersey) Record*, November 20, 2001, a03.

40. J. Gordon, "Our Air, Their Air: Most of It Is Bad," *New York Times*, November 10, 2002, 1; M. Janofsky, "Change to the Clean Air Act Is Built into New Energy Bill," *New York Times*, April 16, 2005.

41. *Chattanooga Times Free Press*, "Tennesseans Can Mold Future of Transportation in State," November 7, 2004, F1.

42. Kiehl, "Cleaner-Fuel Buses Sought to Reduce Air Pollution."

43. J. Barry, "Summer's Bad Air Days Are Upon Us; Health Situation Seen as Acute in Passaic," *Bergen (New Jersey) Record*, June 20, 2003, L1.

44. G. A. Owens, "Power Plants Do Count," *Raleigh News and Observer*, September 8, 2001, A19; J. E. Shiffer, "Populace Cries Out for Clean Air," *Raleigh News and Observer*, October 11, 2000, A3.

45. K. Joy, "Soot Pushes Air Quality to 'Unhealthy'; 4th Alert This Year in Six-County Area Means Breathing Problems for Many," *Columbus Dispatch*, June 17, 2003, 09C.

46. M. Simmons, "Environmentalists Have 'Hot List' Ready," *Knoxville News-Sentinel*, January 22, 2003, B-1.

47. D. Shapley, "Some Fear Boost in Air Pollution," *Poughkeepsie (New York) Journal*, August 29, 2003, 1B.

48. For a more comprehensive list of "some of the worst air pollution" claims, see J. Schwartz, "Air Quality: Much Worse on Paper than in Reality," American Enterprise Institute, May 2005, http://www.aei.org/docLib/20050602_EPOMay_Junenewg %282%29.pdf (accessed September 4, 2006).

49. M. Enge, "Study Links Pollution to Asthma in Children; Active Kids in Smoggy Areas at More Risk, Researchers Say," *San Jose Mercury News*, February 1, 2002, 21A.

50. R. O'Toole, transit market share data table, American Dream Coalition, 2004, http://www.americandreamcoalition.org/transitshare.xls (accessed September 5, 2006).

51. J. Schwartz, *Grading the Graders: How Advocacy Groups' Report Cards Mislead the Public on Air Pollution and Urban Transport*, Reason Public Policy Institute, December 2001, http://www.rppi.org/pu13.html (accessed September 5, 2006).

52. U.S. Public Interest Research Group, *More Highways, More Pollution: Road-Building and Air Pollution in America's Cities* (Washington, D.C.: U.S. PIRG, March 2004).

53. See figure I-1 in the Introduction.

54. U.S. Department of Transportation, Volpe Center, "Trends in Personal Motor Vehicle Ownership and Use: Evidence from the Nationwide Personal Transportation Survey," by D. Pickrell and P. Schmiek, April 23, 1998, http://npts.ornl.gov/npts/1995/Doc/Envecon.pdf (accessed September 5, 2006).

55. Abt Associates, *Particulate-Related Health Impacts of Emissions in 2001 from 41 Major US Power Plants*, November 2002, http://www.environmentalintegrity.org/pubs/PMHealthImpact2001.pdf (accessed November 27, 2006).

56. Sulfate results are from EPA's CASTNET data Web site, http://cfpub.epa.gov/gdm/index.cfm?fuseaction=aciddeposition.wizard (accessed November 27, 2006).

57. Abt Associates, *Particulate-Related Health Impacts of Emissions in 2001 from 41 Major US Power Plants*.

58. U.S. Public Interest Research Group, *Darkening Skies: Trends Toward Increasing Power Plant Emissions* (Washington, D.C.: U.S. PIRG, April 2002); Clean Air Task Force, *Death, Disease and Dirty Power: Mortality and Health Damage Due to Air Pollution from Power Plants*, October 2000, http://www.cleartheair.org/fact/mortality/mortality lowres.pdf; Clean Air Task Force, *Power to Kill: Death and Disease from Power Plants Charged with Violating the Clean Air Act*, July 2001, http://www.cleartheair.org/relatives/18300.pdf#search=%22Power%20to%20Kill%3A%20Death%20and%20Disease%20from%20Power%20Plants%22 (accessed September 11, 2006); U.S. Public Interest Research Group, *Pollution on the Rise: Local Trends in Power Plant Pollution*, January 2005, http://uspirg.org/reports/pollutionontherise.pdf (accessed September 11, 2006).

59. EPA has since adopted the Clean Air Interstate Rule, which requires much larger reductions in power plant SO_2 emissions (see chapter 4). U.S. Environmental Protection Agency, *EPA's Acid Rain Program: Results of Phase I, Outlook for Phase II* (Washington, D.C.: Government Printing Office, 2001).

60. Based on sulfate monitoring data downloaded from EPA's CASTNET data Web site, http://cfpub.epa.gov/gdm/index.cfm?fuseaction=aciddeposition.wizard (accessed November 27, 2006).

61. Clean Air Trust, "With Bush Poised to Weaken Air Enforcement, New Survey Finds Massive Smog Problem in 2002," press release, September 30, 2002, http://www.cleanairtrust.org/release.093002.html (accessed September 11, 2006).

62. Broder, "Cleaner Air in Los Angeles?"

63. D. E. Beeman, "Weather Spares Area Title of 'Smog Capital,'" *Riverside Press Enterprise*, October 7, 2004; Broder, "Cleaner Air in Los Angeles?"; C. Fly, "Another Cool Summer Bodes Well for Air Quality across State," *Tennessean*, October 4, 2004, 2B; P. Henetz, "Storms, Mild Temperatures Help Keep Ozone Levels Low," *Salt Lake Tribune*, October 2, 2004; PIRG, *Danger in the Air*, September 2004, http://www.uspirg.org/uploads/ST/-D/ST-DQWGxwXgNrFoTNz3WyA/Danger-in-the-Air-Final.pdf (accessed November 27, 2006).

64. McFarling and Bustillo, "2005 Vying with '98 as Record Hot Year."

65. American Lung Association, *State of the Air: 2004*.

66. American Lung Association, *State of the Air: 2005*.

67. J. A. Patz, P. L. Kinney, M. L. Bell, et al., *Heat Advisory: How Global Warming Causes More Bad Air Days*, Natural Resources Defense Council, July 2004, http://www.nrdc.org/globalWarming/heatadvisory/heatadvisory.pdf (accessed September 11, 2006).

68. U.S. Department of State, *U.S. Climate Action Report*, May 2002, http://yosemite.epa.gov/oar/globalwarming.nsf/content/ResourceCenterPublicationsUSClimateActionReport.html (accessed September 11, 2006).

69. Attorney General of Massachusetts, "MA Impacts Fact Sheet: Massachusetts Global Warming Impacts as Reported by the US EPA," http://www.ago.state.ma.us/sp.cfm?pageid=1618 (accessed September 11, 2006).

70. K. Knowlton, J. E. Rosenthal, C. Hogrefe, et al., "Assessing Ozone-Related Health Impacts under a Changing Climate," *Environmental Health Perspectives* 112 (2004): 1557–63.

71. National-average temperature data for June–August. National Climatic Data Center, "U.S. Climate at A Glance," http://www.ncdc.noaa.gov/oa/climate/research/cag3/cag3.html (accessed November 27, 2006).

72. Based on a three-year moving average.

73. Attorney General of Massachusetts, "MA Impacts Fact Sheet."

74. For $PM_{2.5}$ composition around the United States, see U.S. Environmental Protection Agency, *Latest Findings on National Air Quality: 2002*.

75. J. Aw and M. J. Kleeman, "Evaluating the First-Order Effect of Inter-Annual Temperature Variability on Urban Air Pollution," *Journal of Geophysical Research—Atmospheres* 108 (2003): 7-1–7-18.

76. NRDC, *Heat Advisory*, 4.

77. We critique these health claims in chapter 7. For now the important point is that regulators and environmentalists consider $PM_{2.5}$ to have substantially greater health impacts than ozone. For EPA's attribution of the health benefits of the Clean Air Act, see U.S. Environmental Protection Agency, *The Benefits and Costs of the Clean Air Act, 1970 to 1990*, October 1997, http://www.epa.gov/air/sect812/; U.S. Environmental Protection Agency, *The Benefits and Costs of the Clean Air Act 1990 to 2010*, EPA Report to Congress, November 1999, http://www.epa.gov/air/sect812/copy 99.html (both accessed September 11, 2006).

Chapter 6: How Many Americans Live in Areas That Violate Federal Air Pollution Standards? Far Fewer Than You Think

1. ALA gives a county a failing ozone grade if, by ALA's counting method, the county averages more than three 8-hour ozone exceedance days per year.

2. American Lung Association, *State of the Air: 2006*.

3. We estimate 48 million based on analysis of ozone levels in counties that have at least one ozone monitor. EPA's and ALA's estimates also include only such counties. We also estimate that roughly another 10 million people lived in areas of unmonitored counties that violated the 8-hour ozone standard. We cite the 48 million figure here because it is directly comparable with EPA's and ALA's estimates. We provide more detail on the derivation of these numbers later in this chapter.

4. This is larger than the 136 million EPA claimed for 2002 and the 100 million it claimed for 2003. The reason EPA's number went up even though ozone went down is that the agency has begun including counties in its tally that are in full attainment of the 8-hour ozone standard but are part of a regulatory nonattainment area; U.S. Environmental Protection Agency, "EPA Issues Designations on Ozone Health Standards," press release, April 15, 2004, http://yosemite.epa.gov/opa/admpress.nsf/ b1ab9f485b098972852562e7004dc686/f2673d2323be58b385256e77005aa 9af?OpenDocument (accessed September 11, 2006); U.S. Environmental Protection Agency, *The Ozone Report: Measuring Progress through 2003*, April 2004, http://www. epa.gov/airtrends/aqtrnd04/pdfs/2003ozonereport.pdf (accessed September 11, 2006).

5. Once again, these are figures for counties with $PM_{2.5}$ monitors, because we are comparing with EPA and ALA figures that are also based only on counties that have $PM_{2.5}$ monitors.

6. ALA gives a county a failing grade if it averages at least 3.3 days per year with ozone at 0.085 ppm or more. This is more stringent than the 8-hour standard in two ways. First, ALA's "countywide" method of counting ozone exceedance days results in more exceedance days than ever actually occur even at the worst monitoring location in the county. Second, violation of the 8-hour standard occurs when the fourth-highest 8-hour ozone reading at any monitoring site in a county averages at least 0.085 ppm for the most recent three-year period. In practice this means that violating areas average about 4 to 5 exceedance days per year, rather than the 3.3 days ALA uses as a cutoff.

7. Based only on monitored counties, the improvements are 53 percent and 33 percent.

8. The estimate of population living in areas that violate a pollution standard was performed by Dennis Kahlbaum of Air Improvement Resource.

9. We don't estimate the number of people in unmonitored counties living in areas that violate either or both ozone or $PM_{2.5}$ standards. We simply don't have any data

that would allow us to make even rough estimates of the degree of spatial overlap between $PM_{2.5}$ and ozone violations in unmonitored counties, or in counties that are monitored for only one of the two pollutants.

10. Recall that violation of a pollution standard depends on the most recent three years of monitoring data. Thus, the violation rate for 2005 depends on data for 2003–5, while the violation rate for 2006 depends on data for 2004–6. Thus, the fact that pollution was lower in 2006 than in 2003 means that there will be a drop from 2005 to 2006 in the number of people living in violating areas.

11. U.S. Environmental Protection Agency, *Fact Sheet: Final Revisions to the National Ambient Air Quality Standards for Particle Pollution (Particulate Matter)*.

12. Among many other examples, see, for example, B. Becker, "Policy Levers for Reducing Emissions from Goods Transport: State Policy Options," State and Territorial Air Pollution Program Administrators/Association of Local Air Pollution Control Officials" (paper, Haagen Smit Symposium, Fifth Annual Meeting, California Air Resources Board, Aptos, Calif., April 18–21, 2005), http://www.westcoast-diesel.org/files/clearinghouse-marine/Haagen-Smit%20Symposium%20Policy%20Levers%20for%20Reducing%20Emissions1.pdf (accessed September 11, 2006); M. Bernstein and D. Whitman, "Smog Alert: The Challenges of Battling Ozone Pollution," *Environment*, October 2005, 10–27; Commuterchoice.gov, "Fast Facts: Best Workplaces for Commuters," http://www.commuterchoice.gov/pdf/sanfran/sffastfacts04_5.pdf (accessed September 11, 2006); J. Eilperin, "Proposals Stiffen Standards on Air; EPA Weighs Lowering Soot Limit," *Washington Post*, July 2, 2005, A04; *Environmental Defense*, "Smog Alert: How Commercial Shipping Is Polluting Our Air," June 2004, http://www.environmentaldefense.org/documents/3811_FactSheet_SmogAlertCommercialShipping.pdf (accessed September 11, 2006); D. Feinstein, "Senator Feinstein Discusses Air Quality, Climate Change on Earth Day 2004," press release, April 22, 2004, http://feinstein.senate.gov/04Releases/r-earthday.htm (accessed September 11, 2006); J. Heilprin, "Trucking Industry Won't Fight Diesel Rules," Associated Press, November 17, 2004; B. Henderson, "Decades Later, Environmental Victories Balanced by Changing Challenges," *Charlotte Observer*, April 22, 2005; C. V. Mathai, "Introduction to Stakeholder Perspectives on the Clean Air Interstate and Clean Air Mercury Rules," *Environmental Manager*, August 2005; Midwest Council of State Governments, *Eye on the Environment and Energy Policy*, April 2004; Natural Resources Defense Council, "Asthma and Air Pollution," June 2005, http://www.nrdc.org/health/effects/fasthma.asp (accessed September 11, 2006); U.S. Public Interest Research Group, *Plagued by Pollution*, January 2006, http://cleanairnow.org/pdfs/plaugedbypollution.pdf (accessed September 11, 2006); E. Sahle-Demessie, "The Challenges of Air Pollution and Residual Risk Assessment," *Journal of Environmental Engineering* 132 (2006): 431–32; *Tulsa World*, "Clear Skies or Dirty Air?" March 26, 2005; N. Willcox, J. Rutkowski, and J. Hubbard, "Time to Speak up for Clean Air," *Philadelphia Inquirer*, March 8, 2006, B02.

Chapter 7: Air Pollution and Health

1. R. McConnell, K. T. Berhane, F. Gilliland, et al., "Asthma in Exercising Children Exposed to Ozone: A Cohort Study," *Lancet* 359 (2002): 386–91.

2. See, for example, W. Booth, "Study: Pollution May Cause Asthma; Illness Affects 9 Million U.S. Children," *Washington Post*, February 1, 2002, A1; C. Bowman, "Asthma's Toll: A New Study Links Children's Sports Activities in Smoggy Areas to the Illness," *Sacramento Bee*, February 1, 2002, A1; Enge, "Study Links Pollution to Asthma in Children"; T. Freemantle, "Asthma Risk for Children Soars with High Ozone Levels—Study," *Houston Chronicle*, February 1, 2002, A1.

3. This lower overall asthma risk is discussed in the peer-reviewed journal article the researchers published on the study. McConnell, Berhane, Gilliland, et al., "Asthma in Exercising Children Exposed to Ozone."

4. Yale Center for Environmental Law and Policy, *The Environmental Deficit: Survey on American Attitudes on the Environment*, May 2004, http://www.yale.edu/envirocenter/report-gw.pdf (accessed December 4, 2006). For example, some 80 percent of New Yorkers rate air pollution as a "very serious" or "somewhat serious" problem, as do 77 percent of Texans and 86 percent of New Jerseyans. When asked about the most serious environmental issue facing California, a 34 percent plurality chose air pollution, with "growth" coming in a distant second at 13 percent. Roughly a third of Californians put air pollution first even in San Diego and the San Francisco Bay Area, where almost everyone lives in areas that meet all federal standards. American Viewpoint, *Recent Texas Statewide Survey Findings Prepared for Public Citizen and the Seed Coalition* (Alexandria, Va.: American Viewpoint, 2002); New York Conservation Education Fund, *Key Findings of a Statewide Survey of New York State Residents on Environmental Issues* (New York: New York League of Conservation Voters, 2001); *Star-Ledger*/Eagleton-Rutgers Poll, "Sprawl: New Jerseyans Dislike the Problems, and the Solutions," Eagleton Institute of Politics, press release, September 29, 2002, http://slerp.rutgers.edu/retrieve.php?id=138-6 (accessed September 13, 2006); Baldassare, *PPIC Statewide Survey: Special Survey on Californians*.

5. Clean Air Task Force, *Death, Disease and Dirty Power*; Natural Resources Defense Council, *Our Children at Risk*, November 1997, http://www.nrdc.org/health/kids/ocar/ocarinx.asp (accessed September 13, 2006); Physicians for Social Responsibility, *Children at Risk: How Air Pollution from Power Plants Threatens the Health of America's Children*, May 2002, http://www.cleartheair.org/fact/children/children_at_risk.pdf (accessed September 13, 2006); U.S. Public Interest Research Group, *Danger in the Air*; U.S. Public Interest Research Group, *Plagued by Pollution*; Sierra Club, *Highway Health Hazards*, July 2004, http://www.sierraclub.org/sprawl/report04_highwayhealth/report.pdf (accessed September 13, 2006).

6. Johns Hopkins School of Public Health News Center, "Traffic Exhaust Poisons Home Air," August 31, 1999, http://www.jhsph.edu/PublicHealthNews/Press_Releases/PR_1999/traffic_exhaust.html (accessed September 13, 2006); A. Di Rado, "USC Study Shows Air Pollution May Trigger Asthma in Young Athletes," University

of Southern California, February 1, 2002, http://www.usc.edu/hsc/info/pr/1vol8/803/air.html (accessed September 13, 2006); A. Di Rado, "Smog May Cause Lifelong Lung Deficits," University of Southern California, September 8, 2004, http://www.usc.edu/uscnews/stories/10495.html (accessed September 13, 2006); U.S. Department of Health and Human Services, National Institute of Environmental Health Sciences, "Link Strengthened between Lung Cancer, Heart Deaths and Tiny Particles of Soot," press release, March 5, 2002, http://www.niehs.nih.gov/oc/news/lchlink.htm (accessed September 13, 2006).

7. T. Avril, "Air Pollution's Threat Proving Worse Than Believed," *Philadelphia Inquirer*, November 17, 2004, A01; M. Cone, "State's Air Is Among Nation's Most Toxic," *Los Angeles Times*, March 22, 2006; M. Cone, "Study Finds Smog Raises Death Rate," *Los Angeles Times*, November 17, 2004, A20; Freemantle, "Asthma Risk for Children Soars"; T. Webber, "Don't Breathe Deeply; Heat, Sun Cook Up 1st Ozone Alert in 2 Years," *Indianapolis Star*, June 23, 2005, 1A.

8. California Air Resources Board, *Review of the California Ambient Air Quality Standard for Ozone*, March 17, 2005, http://www.arb.ca.gov/research/aaqs/ozone-rs/ozone-rs.htm (accessed December 5, 2006).

9. J. Schwartz, *Rethinking the California Air Resources Board's Ozone Standards*, American Enterprise Institute, September 2005, http://www.aei.org/doclib/20050912_Schwartzwhitepaper.pdf (accessed September 13, 2006); California Air Resources Board, *Particulate Air Pollution and Morbidity in the California Central Valley: A High Particulate Pollution Region*, by S. F. van den Eeden, C. P. Quesenberry, J. Shan, and F. W. Lurmann (Sacramento: California Air Resources Board, July 2002).

10. U.S. Environmental Protection Agency, *Air Quality Criteria for Ozone and Related Photochemical Oxidants (Second External Review Draft), Volumes I–III*, August 2005, http://www.epa.gov/ttn/naaqs/standards/ozone/s_o3_cr_cd.html (accessed September 13, 2006); J. Schwartz, "Comments on EPA's Ozone Criteria Document," American Enterprise Institute, December 2, 2005, http://www.joelschwartz.com/pdfs/AEI_O3AQCD2.pdf (accessed September 13, 2006).

11. W. J. Gauderman, E. Avol, F. Gilliland, et al., "The Effect of Air Pollution on Lung Development from 10 to 18 Years of Age," *New England Journal of Medicine* 351 (2004): 1057–67.

12. University of Southern California/National Institutes of Health, "Teens in Smoggy Areas at High Risk for Starting Adulthood with Serious Lung Deficit," September 8, 2004, http://www-apps.niehs.nih.gov/centers/Public/news/nws541.htm (accessed November 27, 2006).

13. C. A. Pope III, M. J. Thun, M. M. Namboodiri, et al., "Particulate Air Pollution as a Predictor of Mortality in a Prospective Study of U.S. Adults," *American Journal of Respiratory and Critical Care Medicine* 151 (1995): 669–74.

14. D. Krewski, R. T. Burnett, M. S. Goldberg, et al., *Reanalysis of the Harvard Six Cities Study and the American Cancer Society Study of Particulate Air Pollution and Mortality* (Cambridge, Mass.: Health Effects Institute, July 2000).

15. Ibid; C. A. Pope III, R. T. Burnett, M. J. Thun, et al., "Lung Cancer, Cardiopulmonary Mortality, and Long-Term Exposure to Fine Particulate Air Pollution," *Journal of the American Medical Association* 287 (2002): 1132–41.

16. F. W. Lipfert, H. M. Perry, J. P. Miller, et al., "The Washington University–EPRI Veterans' Cohort Mortality Study," *Inhalation Toxicology* 12, supp. 4 (2000): 41–73.

17. Clean Air Task Force, *Death, Disease and Dirty Power*; Clean Air Task Force, *Power to Kill*; Physicians for Social Responsibility, *Children at Risk*; U.S. Public Interest Research Group, *Danger in the Air*.

18. M. A. Sackner, D. Ford, and R. Fernandez, "Effect of Sulfate Aerosols on Cardiopulmonary Function of Normal Humans," *American Review of Respiratory Diseases* 115 (1977): 240; M. J. Utell, P. E. Morrow, D. M. Speers, et al., "Airway Responses to Sulfate and Sulfuric Acid Aerosols in Asthmatics. An Exposure-Response Relationship," *American Review of Respiratory Disease* 128 (1983): 444–50.

19. J. Q. Koenig, K. Dumler, V. Rebolledo, et al., "Respiratory Effects of Inhaled Sulfuric Acid on Senior Asthmatics and Nonasthmatics," *Archives of Environmental Health* 48 (1993): 171–75.

20. M. T. Kleinman, W. S. Linn, R. M. Bailey, et al., "Effect of Ammonium Nitrate Aerosol on Human Respiratory Function and Symptoms," *Environmental Research* 21 (1980): 317–26; M. J. Utell, A. J. Swinburne, R. W. Hyde, et al., "Airway Reactivity to Nitrates in Normal and Mild Asthmatic Subjects," *Journal of Applied Physiology* 46 (1979): 189–96.

21. American Lung Association, "Medical Journal Watch," http://www.cleanair standards.org/category/medical-journal-watch/. (accessed November 27, 2006).

22. For example, Medical Journal Watch does not include information on any studies by Fred Lipfert, Suresh Moolgavkar, Richard Smith, Gary Koop, William Keatinge, Laura Green, or James Enstrom, all of whom have provided evidence against a connection between low-level air pollution and risk of death (based on a search of the Medical Journal Watch Web site on April 6, 2006, for articles authored by the researchers listed above).

23. This discussion of the implications of HRT studies for air pollution epidemiology is summarized from S. H. Moolgavkar, "A Review and Critique of the EPA's Rationale for a Fine Particle Standard," *Regulatory Toxicology and Pharmacology* 42 (2005): 123–44.

24. M. J. Stampfer and G. A. Colditz, "Estrogen Replacement Therapy and Coronary Heart Disease: A Quantitative Assessment of the Epidemiologic Evidence," *Preventive Medicine* 20 (1991): 47–63.

25. E. Barrett-Connor, "Clinical Review 162: Cardiovascular Endocrinology 3: An Epidemiologist Looks at Hormones and Heart Disease in Women," *Journal of Clinical Endocrinology and Metabolism* 88 (2003): 4031–42; J. E. Rossouw, G. L. Anderson, R. L. Prentice, et al., "Risks and Benefits of Estrogen Plus Progestin in Healthy Postmenopausal Women: Principal Results from the Women's Health Initiative Randomized Controlled Trial," *Journal of the American Medical Association* 288 (2002): 321–33.

26. J. P. Ioannidis, "Why Most Published Research Findings Are False," *PLoS Medicine* 2 (2005): e124, http://medicine.plosjournals.org/archive/1549-1676/2/8/pdf/10.1371_journal.pmed.0020124-L.pdf (accessed March 28, 2007).

27. S. A. Beresford, K. C. Johnson, C. Ritenbaugh, et al., "Low-Fat Dietary Pattern and Risk of Colorectal Cancer: The Women's Health Initiative Randomized Controlled Dietary Modification Trial," *Journal of the American Medical Association* 295 (2006): 643–54; B. V. Howard, L. Van Horn, J. Hsia, et al., "Low-Fat Dietary Pattern and Risk of Cardiovascular Disease: The Women's Health Initiative Randomized Controlled Dietary Modification Trial," *Journal of the American Medical Association* 295 (2006): 655–66; R. L. Prentice, B. Caan, R. T. Chlebowski, et al., "Low-Fat Dietary Pattern and Risk of Invasive Breast Cancer: The Women's Health Initiative Randomized Controlled Dietary Modification Trial," *Journal of the American Medical Association* 295 (2006): 629–42.

28. G. Kolata, "Big Study Finds No Clear Benefit of Calcium Pills," *New York Times*, February 16, 2006.

29. G. D. Smith, "Reflections on the Limitations to Epidemiology," *Journal of Clinical Epidemiology* 54 (2001): 325–31.

30. Publication bias is a well-documented problem in a range of disciplines. See, for example, V. M. Montori, M. Smieja, and G. H. Guyatt, "Publication Bias: A Brief Review for Clinicians," *Mayo Clinic Proceedings* 75 (2000): 1284–88; A. Thornton and P. Lee, "Publication Bias in Meta-Analysis: Its Causes and Consequences," *Journal of Clinical Epidemiology* 53 (2000): 207–16.

31. H. Anderson, R. Atkinson, J. Peacock, L. Marston, and K. Konstantinou, *Meta-Analysis of Time-Series Studies and Panel Studies of Particulate Matter (PM) and Ozone*, World Health Organization, 2004, www.euro.who.int/document/e82792.pdf (accessed September 13, 2006).

32. T. Lumley and L. Sheppard, "Time Series Analyses of Air Pollution and Health: Straining at Gnats and Swallowing Camels?" *Epidemiology* 14 (2003): 13–14.

33. Ioannidis, "Why Most Published Research Findings Are False"; Smith, "Reflections on the Limitations to Epidemiology"; S. Begley, "New Journals Bet 'Negative Results' Save Time, Money," *Wall Street Journal*, September 15, 2006, B1; Taubes, "Epidemiology Faces Its Limits"; P. C. Austin, M. M. Mamdani, D. N. Juurlink, and J. E. Hux, "Testing Multiple Statistical Hypotheses Resulted in Spurious Associations: A Study of Astrological Signs and Health," *Journal of Clinical Epidemiology* 59 (2006): 964–69.

34. Begley, "New Journals Bet 'Negative Results' Save Time, Money."

35. American Association for the Advancement of Science, Annual Meeting, Program and Events, http://www.aaas.org/meetings/Annual_Meeting/02_PE/pe_chron_070216fri.shtml (accessed March 21, 2007).

36. Goklany, *Clearing the Air*.

37. Among many examples, see R. Wilson and J. Spengler, *Particles in Our Air: Concentrations and Health Effects* (Cambridge, Mass.: Harvard University Press, 1996).

Also see Natural Resources Defense Council, "Particulate Pollution," May 7, 1996, http://www.nrdc.org/air/pollution/qbreath.asp (accessed September 15, 2006).

38. As of this writing, the most recent critique of the $PM_{2.5}$ mortality literature is Moolgavkar, "A Review and Critique."

39. D. W. Dockery, C. A. Pope III, X. Xu, et al., "An Association Between Air Pollution and Mortality in Six U.S. Cities," *New England Journal of Medicine* 329 (1993): 1753–59; Pope, Thun, Namboodiri, et al., "Particulate Air Pollution as a Predictor of Mortality."

40. Krewski, Burnett, Goldberg, et al., *Reanalysis of the Harvard Six Cities Study*; Pope, Burnett, Thun, et al., "Lung Cancer, Cardiopulmonary Mortality, and Long-Term Exposure to Fine Particulate Air Pollution."

41. To calculate the size of the PM mortality relationship for 1990–98, you need to take into account not only the effect sizes for 1982–89 and 1982–98, but also the fact that the number of deaths in the cohort was twice as high for the 1990–98 period as for the 1982–89 period (because the cohort was older), and that the size of the cohort was expanded by 22 percent by including additional cities. Assuming the estimated $PM_{2.5}$ risks were the same, on average, for the added cities as for the original ones, the association of $PM_{2.5}$ with mortality would have dropped 73 percent between the two periods.

42. Summarized in table 10 of J. E. Enstrom, "Fine Particulate Air Pollution and Total Mortality Among Elderly Californians, 1973–2002," *Inhalation Toxicology* 17 (2005): 803–16.

43. Krewski, Burnett, Goldberg, et al., *Reanalysis of the Harvard Six Cities Study*.

44. Ibid.

45. Lipfert, Perry, Miller, et al., "The Washington University–EPRI Veterans' Cohort Mortality Study."

46. F. W. Lipfert and S. C. Morris, "Temporal and Spatial Relations Between Age Specific Mortality and Ambient Air Quality in the United States: Regression Results for Counties, 1960–97," *Occupational and Environmental Medicine* 59 (2002): 156–74.

47. Enstrom, "Fine Particulate Air Pollution and Total Mortality."

48. American Lung Association, "Long-Term Epidemiological Studies Are Validated in Independent Reanalysis," October 10, 2001, http://www.cleanairstandards.org/article/2001/10/32 (accessed December 5, 2006).

49. See, for example, Health Effects Institute, *Revised Analyses of Time-Series Studies of Air Pollution and Health* (Boston: Health Effects Institute, May 2003).

50. M. L. Bell, F. Dominici, and J. M. Samet, "A Meta-Analysis of Time-Series Studies of Ozone and Mortality with Comparison to the National Morbidity, Mortality, and Air Pollution Study," *Epidemiology* 16 (2005): 436–45; M. L. Bell, A. McDermott, S. L. Zeger, et al., "Ozone and Short-Term Mortality in 95 US Urban Communities, 1987–2000," *Journal of the American Medical Association* 292 (2004): 2372–78; K. Ito, S. F. De Leon, and M. Lippmann, "Associations Between Ozone and

Daily Mortality: Analysis and Meta-Analysis," *Epidemiology* 16 (2005): 446–57; J. I. Levy, S. M. Chemerynski, and J. A. Sarnat, "Ozone Exposure and Mortality: An Empiric Bayes Metaregression Analysis," *Epidemiology* 16 (2005): 458–68.

51. T. Lumley and L. Sheppard, "Time Series Analyses of Air Pollution and Health: Straining at Gnats and Swallowing Camels?" *Epidemiology* 14 (2003): 13–14.

52. See, for example, M. Clyde, "Model Uncertainty and Health Effect Studies for Particulate Matter," *Environmetrics* 11 (2000): 745–63; S. H. Moolgavkar, *Review of Chapter 8 of the Criteria Document for Particulate Matter* (Comments Submitted to EPA) (Alexandria, Va.: Sciences International, 2002); S. H. Moolgavkar and E. G. Luebeck, "A Critical Review of the Evidence on Particulate Air Pollution and Mortality," *Epidemiology* 7 (1996): 420–28; S. H. Moolgavkar, E. G. Luebeck, T. A. Hall, and E. L. Anderson, "Particulate Air Pollution, Sulfur Dioxide, and Daily Mortality: A Reanalysis of the Steubenville Data," *Inhalation Toxicology* 7 (1995): 35–44; R. L. Smith, J. M. Davis, et al., "Regression Models for Air Pollution and Daily Mortality: Analysis of Data from Birmingham, Alabama," *Environmetrics* 11 (2000): 719–43; J. M. Samet, F. Dominici, A. McDermott, and S. L. Zeger, "New Problems for an Old Design: Time Series Analyses of Air Pollution and Health," *Epidemiology* 14 (2003): 11–12; R. Klemm "Reanalysis of Harvard Six-City Mortality Study Replication," EPA Workshop on GAM-Related Statistical Issues in PM Epidemiology, Durham, North Carolina, November 4–6, 2002.

53. Smith, Davis, Sacks, et al., "Regression Models for Air Pollution and Daily Mortality."

54. M. Bell, J. Samet, and F. Dominici, *Ozone and Mortality: A Meta-Analysis of Time-Series Studies and Comparison to a Multi-City Study* (The National Morbidity, Mortality, and Air Pollution Study), Johns Hopkins School of Public Health, July 19, 2004, http://www.bepress.com/cgi/viewcontent.cgi?article=1057&context=jhubiostat (accessed September 15, 2006).

55. Anderson, Atkinson, Peacock, et al., *Meta-Analysis of Time-Series Studies and Panel Studies of Particulate Matter (PM) and Ozone.*

56. Bell, McDermott, Zeger, et al., "Ozone and Short-Term Mortality in 95 US Urban Communities, 1987–2000." See figure 3, 2376.

57. Moolgavkar, *Review of Chapter 8 of the Criteria Document for Particulate Matter.*

58. Bell, Dominici, and Samet, "A Meta-Analysis of Time-Series Studies of Ozone and Mortality"; Ito, De Leon, and Lippmann, "Associations Between Ozone and Daily Mortality"; Levy, Chemerynski, and Sarnat, "Ozone Exposure and Mortality." The updated ozone staff report has not been posted on CARB's Web site as of this writing, but was emailed to CARB's "ozone standard stakeholders" list on July 18, 2005, and the updated version is dated July 18, 2005.

59. Goodman, "The Methodologic Ozone Effect." 431.

60. The various models differed in the way they controlled for weather and long-term trends. K. Ito, "Associations of Particulate Matter Components with Daily Mortality and Morbidity in Detroit," in *Revised Analyses of Time-Series Studies of Air*

Pollution and Health (Boston: Health Effects Institute, May 2003), http://pubs.healtheffects.org/getfile.php?u=21 (accessed November 27, 2006).

61. G. Koop and L. Tole, "Measuring the Health Effects of Air Pollution: To What Extent Can We Really Say That People Are Dying from Bad Air?" *Journal of Environmental Economics and Management* 47 (2004): 30–54.

62. Even the BMA study described above did not account for this additional weather effect. W. R. Keatinge and G. C. Donaldson, "Heat Acclimatization and Sunshine Cause False Indications of Mortality Due to Ozone," *Environmental Research* 100 (2006): 387–93.

63. L. Green, E. Crouch, M. Ames, and T. Lash, "What's Wrong with the National Ambient Air Quality Standard (NAAQS) for Fine Particulate Matter (PM2.5)?" *Regulatory Toxicology and Pharmacology* 35 (2002): 327; L. C. Green and S. R. Armstrong, "Particulate Matter in Ambient Air and Mortality: Toxicologic Perspectives," *Regulatory Toxicology and Pharmacology* 38 (2003): 326–35; Moolgavkar, "A Review and Critique."

64. Green and Armstrong, "Particulate Matter in Ambient Air and Mortality."

65. Q. Sun, A. Wang, X. Jin, et al., "Long-Term Air Pollution Exposure and Acceleration of Atherosclerosis and Vascular Inflammation in an Animal Model," *Journal of the American Medical Association* 294 (2005): 3003–10.

66. U.S. Department of Health and Human Services, National Institute of Environmental Health Sciences, "Air Pollution, High-Fat Diet Cause Atherosclerosis in Laboratory Mice," press release, December 22, 2005, http://www.nih.gov/news/pr/dec2005/niehs-22.htm (accessed September 15, 2006).

67. Newspapers carrying articles on the study included the *Los Angeles Times*, *Houston Chronicle*, *Philadelphia Inquirer*, and several others.

68. A. S. Plump, J. D. Smith, T. Hayek, et al., "Severe Hypercholesterolemia and Atherosclerosis in Apolipoprotein E-Deficient Mice Created by Homologous Recombination in Es Cells," *Cell* 71 (1992): 343–53; S. H. Zhang, R. L. Reddick, J. A. Piedrahita, and M. Maeda, "Spontaneous Hypercholesterolemia and Arterial Lesions in Mice Lacking Apolipoprotein E," *Science* 258 (1992): 468–71.

69. See table 70 in National Center for Health Statistics, *Health, United States, 2005*, http://www.cdc.gov/nchs/data/hus/hus05.pdf#070 (accessed September 15, 2006).

70. Based on National Health and Nutrition Examination Survey (NHANES) data on 4,090 adult men collected 1999–2002. Data were downloaded from U.S. Department of Health and Human Services, National Center for Health Statistics, *National Health and Nutrition Examination Survey*, August 16, 2006, http://www.cdc.gov/nchs/nhanes.htm (accessed September 21, 2006).

71. M. Bustillo and M. Cone, "EPA Issues New Plan to Limit Soot; Critics Say the Revised Standard Is Too Weak to Properly Protect the Public from Health Dangers Caused by Breathing Particulates," *Los Angeles Times*, December 21, 2005.

72. U.S. Department of Health and Human Services, National Institute of Environmental Health Sciences, "Air Pollution, High-Fat Diet Cause Atherosclerosis in Laboratory Mice."

73. CARB collects hourly PM$_{2.5}$ readings in several areas of California, including Modesto and Riverside. We used the most recent twelve-month period of data available (February 1, 2005, to January 31, 2006) at the time we did this analysis in March 2006. Data source: California Air Resources Board, "Air Quality Information Page: AQMIS2 Real-Time Query Tool," http://www.arb.ca.gov/aqmis2/paqdselect.php (accessed September 21, 2006).

74. Based on the data discussed in the previous note.

75. U.S. Department of Health and Human Services, National Institute of Environmental Health Sciences, "Particulate Air Pollution and a High Fat Diet: A Potentially Deadly Combination," by M. Lippmann, L. C. Chen, and Rajagopalan, 2005, www.niehs.nih.gov/dert/profiles/hilites/2005/pm-diet.htm (accessed September 15, 2006).

76. U.S. Environmental Protection Agency, *The Benefits and Costs of the Clean Air Act 1990 to 2010*; U.S. Environmental Protection Agency, *2003–2008 EPA Strategic Plan* (Washington, D.C.: September 30, 2003); U.S. Environmental Protection Agency, *The Benefits and Costs of the Clean Air Act, 1970 to 1990*.

77. The Centers for Disease Control (CDC) estimates asthma prevalence trends through its annual National Health Interview Survey. The CDC changed its asthma survey questions in 1997, preventing comparison with data collected up to 1996. Between 1997 and 2000, the CDC stopped asking people whether they currently had asthma. However, in 1997 CDC began asking people who had ever been diagnosed with asthma whether they had had an attack in the past twelve months. In 2001, CDC began once again to ask people whether they currently had asthma, but with a slightly different question than the pre-1997 surveys. Based on these data, the prevalence of asthma attacks leveled off from 1997 to 2003, while the prevalence of asthma declined from 2001 to 2003. American Lung Association, *Trends in Asthma Morbidity and Mortality*, May 2005, http://www.lungusa.org/atf/cf/%7B7A8D42C2-FCCA-4604-8ADE-7F5D5E762256%7D/ASTHMA1.PDF (accessed September 15, 2006); D. M. Mannino, D. M. Noma, L. J. Akinbami, and J. E. Moorman, "Surveillance for Asthma—United States, 1980–1999," *Morbidity and Mortality Weekly Report* 51, no. SS01 (2002): 1–13.

78. Trends in air pollution levels in California were determined from monitoring data retrieved from California Air Resources Board, *2003 Air Quality Data CD*.

79. See, for example, *Fresno Bee*, "Asthma in the Valley; More Research Is Needed into a Disease That Runs Rampant Here," October 4, 2004; Natural Resources Defense Council, "EPA Set to Launch New Study on Causes of Asthma," last updated October 31, 2002, http://www.nrdc.org/bushrecord/articles/br_1157.asp?t=t (accessed September 15, 2006); *Sacramento Bee*, "Smog and Asthma: The Link—and Threat—Are Real," May 6, 2003, B6; R. Sanchez, "In Calif., a Crackling Controversy over Smog; Illnesses Drive Push to Ban Fireplaces," *Washington Post*, February 16, 2003, A1; D. S. Stanley, "Stop the Spread of Asthma by Cleaning Up Our Air," *Fresno Bee*, August 7, 2004, B9; Surface Transportation Policy Project, *Clearing the Air* (Washington, D.C., August 2003).

80. For CARB's press release, see California Air Resources Board, "Study Links Air Pollution and Asthma," press release, January 31, 2002, http://www.arb.ca.gov/newsrel/nr013102.htm.

81. This result is discussed in the peer-reviewed journal article the researchers published on the study. McConnell, Berhane, Gilliland, et al., "Asthma in Exercising Children."

82. The top of the 95 percent confidence interval for relative risk was 1.0, making the result just a hair short of statistical significance.

83. Pollution monitoring data from the Children's Health Study were provided by CARB's staff.

84. California Air Resources Board, "Study Links Air Pollution and Asthma."

85. Norman Edelman, quoted in S. Borenstein, "Air Pollution Is a Cause of Asthma, Study Contends," *Philadelphia Inquirer*, February 1, 2002, A04.

86. Freemantle, "Asthma Risk for Children Soars"

87. In fact, even the worst areas of Sacramento never average more than a few days per year in exceedance of the 1-hour ozone standard and twenty or so days per year exceeding the 8-hour standard—ozone levels typical of the "medium-ozone" CHS communities, in which there was no relationship between air pollution and asthma risk; Jesse Joad, quoted in Bowman, "Asthma's Toll."

88. G. D. Thurston and D. V. Bates, "Air Pollution as an Underappreciated Cause of Asthma Symptoms," *Journal of the American Medical Association* 290 (2003): 1915–17.

89. Surface Transportation Policy Project, *Clearing the Air.*

90. For international data, see "Worldwide Variation in Prevalence of Symptoms of Asthma, Allergic Rhinoconjunctivitis, and Atopic Eczema: ISAAC. The International Study of Asthma and Allergies in Childhood (ISAAC) Steering Committee," *Lancet* 351 (1998): 1225–32.

91. J. Heinrich, B. Hoelscher, C. Frye, et al., "Trends in Prevalence of Atopic Diseases and Allergic Sensitization in Children in Eastern Germany," *European Respiratory Journal* 19 (2002): 1040–6; J. Heinrich, K. Richter, H. Magnussen, and H. E. Wichmann, "Is the Prevalence of Atopic Diseases in East and West Germany Already Converging?" *European Journal of Epidemiology* 14 (1998): 239–45.

92. J. Gauderman, E. Avol, F. Lurmann, et al., "Childhood Asthma and Exposure to Traffic and Nitrogen Dioxide," *Epidemiology* 16 (2005): 737–43.

93. These are the ranges from the twenty-fifth to the seventy-fifth percentiles for NO_2 and distance from a freeway. As you might expect, the two measures of exposure to traffic pollution were correlated, with a coefficient of –0.54. That is, greater distance from a freeway corresponded to lower ambient NO_2 levels near the home.

94. A. W. Gertler, M. Abu-Allaban, W. Coulombe, et al., "Measurements of Mobile Source Particulate Emissions in a Highway Tunnel," *International Journal of Vehicle Design* 27 (2002): 86–93.

95. Kirchstetter, Hooper, Apte, et al., "Characterization of Particle and Gas Phase Pollut-ant Emissions from Heavy- and Light-Duty Vehicles in a California Roadway Tunnel."

96. California Department of Transportation, Division of Transportation System Information, *Truck Kilometers of Travel, California State Highway System, 1986–2001*, 2003, http://www.dot.ca.gov/hq/tsip/otfa/mtab/Trucks/TKT2001.pdf (accessed September 15, 2006). This is of course a statewide average, but it provides a rough estimate of the increase in truck miles of travel over time.

97. Sierra Club, *Highway Health Hazards.*

98. Mark Brown, quoted in N. Strassman, "Tarrant 19th on List of Poor Air Quality," *Fort Worth Star-Telegram*, May 1, 2003.

99. Carolinas Clean Air Coalition, "Impacts of Ozone on Our Health," http://www.clean-air-coalition.org/Air%20Basics/ozone_impact.htm (accessed November 27, 2006).

100. N. Kunzli, F. Lurmann, M. Segal, et al., "Association between Lifetime Ambient Ozone Exposure and Pulmonary Function in College Freshmen—Results of a Pilot Study," *Environmental Research* 72 (1997): 8–23.

101. American Lung Association, *State of the Air: 2003.*

102. American Lung Association of California, *Recent Scientific Findings on Health Effects of Air Pollution and Diesel Exhaust*, 2003, http://www.californialung.org/spotlight/cleanair03_research.html (accessed September 15, 2006).

103. A. Fell, "Primate Research Shows Link between Ozone Pollution, Asthma," October 13, 2000, http://www-dateline.ucdavis.edu/101300/DL_asthma.html (accessed September 15, 2006).

104. Gauderman, Avol, Gilliland, et al., "The Effect of Air Pollution on Lung Development."

105. The CHS study set up special-purpose monitors to measure pollution levels in the communities where the study was performed. CARB staff provided us with the data from these monitors.

106. Carolinas Clean Air Coalition, "Impacts of Ozone on Our Health."

107. The actual range reported in the study was 5 to 28 $\mu g/m^3$. However, $PM_{2.5}$ was measured using a different method from the one EPA began requiring in 1999 to determine compliance with the federal standard. The CHS measured two-week-average $PM_{2.5}$, which understates levels because it allows some "semivolatile" species to evaporate, both because of the long collection time and because the filters are at ambient temperature. The new federal method measures daily-average $PM_{2.5}$ and keeps the filters cooled to prevent evaporation. Because we're comparing the CHS measurements with ones from around the country using the new EPA method, we've corrected the former to make them equivalent to the latter. CARB's research staff provides details on the difference in readings between old and new $PM_{2.5}$ monitors in Motallebi, Taylor, Clinton, Croes, et al., "Particulate Matter in California: Part 1."

108. See figure 3 in Gauderman, Avol, Gilliland, et al., "The Effect of Air Pollution on Lung Development."

109. Based on IPN data for Riverside collected in the early 1980s, and $PM_{2.5}$ data collected by CARB in 1988 and 1989 and retrieved from California Air Resources Board, *2006 Air Quality Data CD*. As discussed in chapter 3, the Federal Reference

Method $PM_{2.5}$ samplers EPA has required since 1999 read about 14 percent higher, for any given actual ambient $PM_{2.5}$ level, when compared with these earlier data, which were collected with dichotomous samplers. Thus, we have applied this 14 percent correction factor as detailed by CARB's research staff in Motallebi, Taylor, Clinton, Croes, et al., "Particulate Matter in California: Part 1."

110. Di Rado, "Smog May Cause Lifelong Lung Deficits."

111. Kenneth Olden, quoted in U.S. Department of Health and Human Services, National Institute of Environmental Health Sciences, "New Research Shows Air Pollution Can Reduce Children's Lung Function," press release, September 9, 2004, http://www.nih.gov/news/pr/sep2004/niehs-08a.htm (accessed September 15, 2006).

112. Here's how: In Gauderman, Avol, Gilliland, et al., "The Effect of Air Pollution on Lung Development," note first from table 3 that $PM_{2.5}$ was associated with a 79.7 milliliter (ml) reduction in FEV_1 between the least and most polluted communities. Then, from table 2, note that average FEV_1 was 3,332 ml for girls and 4,464 ml for boys at eighteen years of age. Given that there were 876 girls and 883 boys in the study (see p. 1059, column 1), the weighted average FEV_1 for the study population was 3,900 ml. The percentage decline is then 79.7/3,900 = 0.02, or 2 percent.

113. The researchers used a regression model to generate this predicted value.

114. This is assuming the "predicted" lung capacity values are valid. The *NEJM* paper provides few details on the model or the underlying distribution of lung-function test scores by community. Thus, another problem with this outcome measure is that it depends on something that wasn't actually measured!

115. U.S. Department of Health and Human Services, National Institute of Environmental Health Sciences, "New Research Shows Air Pollution Can Reduce Children's Lung Function."

116. American Lung Association, *State of the Air: 2005*.

117. N. Bryant, "What Air Quality Problem?" *Charlotte Observer*, September 1, 2005, 10A. We asked Bryant, executive director of the Carolinas Clean Air Coalition, to send us her source for the claim of a 20 percent reduction in lung function. She sent a copy of the NIH press release discussed here (emails between Joel Schwartz and Nancy Bryant, September 2004, on file with the authors). National Institutes of Health, National Institute of Environmental Health Sciences, press release, "New Research Shows Air Pollution Can Reduce Children's Lung Function," September 9, 2004, http://www.nih.gov/news/pr/sep2004/niehs-08a.htm (accessed November 27, 2006).

118. Lori Kobza-Lee, spokeswoman for Sacramento Metropolitan Air Quality Management District, quoted in H. Gomez, "Mercury's Up, Air Quality's Down; Even the Bay Area Swelters as the Heat Overpowers Its Typical Breezes," *Sacramento Bee*, June 27, 2003.

119. U.S. Environmental Protection Agency, "National Ambient Air Quality Standards for Ozone: Proposed Decision," *Federal Register* 61, no. 241 (December 13, 2006):65715–50, http://www.epa.gov/fedrgstr/EPA-AIR/1996/December/Day-13/pr-23901.txt.html (accessed December 5, 2006).

120. This analysis assumes that ozone causes harm only at levels above the 8-hour standard. Benefits will be a few times greater if they continue to accrue when ozone is reduced below the level of the standard. The reason for this is that achieving attainment in a given nonattainment region requires reducing ozone below the standard on the worst day at the worst location in that region. But within any given region, ozone already does not exceed the standard on most days in most locations. Nevertheless, the measures necessary to attain the standard on the worst day at the worst location would also reduce ozone on other days and other locations. As a result, most of the reduction in ozone exposure occurs on days and in locations in which ozone already complies with the standard. B. J. Hubbell, A. Hallberg, D. R. McCubbin, and E. Post, "Health-Related Benefits of Attaining the 8-Hr Ozone Standard," *Environmental Health Perspectives* 113 (2005): 73–82.

121. Even the current federal 8-hour standard might not be attainable in some areas of California. CARB's new California ozone standard is substantially more stringent than the federal standard. California Air Resources Board, *Review of the California Ambient Air Quality Standard for Ozone*; R. Lutter, *Is EPA's Ozone Standard Feasible?* AEI-Brookings Joint Center for Regulatory Studies, December 1999, http://www.aei-brookings.org/ publications/ abstract.php?pid=29 (accessed December 5, 2006); Winner and Cass, "Effect of Emissions Control on the Long-Term Frequency Distribution."

122. This assumes that benefits continue to accrue only until ozone levels are reduced to the 0.070 ppm California standard. If they continue to accrue for levels below the standard, then the percentage reduction in total health effects will be about five times greater—1.8 percent for asthma ER visits and 1.2 percent for respiratory hospital admissions. California Air Resources Board, *Review of the California Ambient Air Quality Standard for Ozone*; Schwartz, *Rethinking the California Air Resources Board's Ozone Standards.*

123. California Air Resources Board, *Review of the California Ambient Air Quality Standard for Ozone*; Schwartz, *Rethinking the California Air Resources Board's Ozone Standards.*

124. Hubbell, Hallberg, McCubbin, et al., "Health-Related Benefits of Attaining the 8-Hr Ozone Standard."

125. CARB estimates this number would rise to 4,200 if benefits continue to accrue when ozone continues to be reduced below the 0.070 ppm standard.

126. California Air Resources Board, *Review of the California Ambient Air Quality Standard for Ozone*; Hubbell, Hallberg, McCubbin, et al., "Health-Related Benefits of Attaining the 8-Hr Ozone Standard"; Schwartz, *Rethinking the California Air Resources Board's Ozone Standards.*

127. Clean Air Task Force, *Death, Disease and Dirty Power*; Clean Air Task Force, *Power to Kill.*

128. Abt Associates, *The Particulate-Related Health Benefits of Reducing Power Plant Emissions*, Clean Air Task Force, October 2000, http://cta.policy.net/fact/mortality/ mortalityabt.pdf (accessed September 18, 2006).

129. Ibid., 6-2.

130. U.S. Environmental Protection Agency, *Latest Findings on National Air Quality, 2002 Status and Trends.*

131. U.S. Environmental Protection Agency, *The Particle Pollution Report.*

132. Green and Armstrong, "Particulate Matter in Ambient Air and Mortality"; Kleinman, Linn, Bailey, et al., "Effect of Ammonium Nitrate Aerosol"; Sackner, Ford, and Fernandez, "Effect of Sulfate Aerosols"; Utell, Morrow, Speers, et al., "Airway Responses to Sulfate and Sulfuric Acid Aerosols"; Utell, Swinburne, Hyde, et al., "Airway Reactivity to Nitrates."

133. Koenig, Dumler, Rebolledo, et al., "Respiratory Effects of Inhaled Sulfuric Acid."

134. L. J. Nannini Jr. and D. Hofer, "Effect of Inhaled Magnesium Sulfate on Sodium Metabisulfite-Induced Bronchoconstriction in Asthma," *Chest* 111 (1997): 858–61.

135. California Air Resources Board, *Particulate Air Pollution and Morbidity in the California Central Valley.*

136. California Air Resources Board, "Hospitalizations and Emergency Room Visits Increase Following High Particulate Matter Episodes, Study Finds," press release, February 24, 2003, http://www.arb.ca.gov/newsrel/nr022403.htm (accessed September 18, 2006).

137. U.S. Environmental Protection Agency, *Air Quality Criteria for Ozone*; J. Schwartz, *Comments on EPA's Ozone Criteria Document.*

138. For additional examples, see Schwartz, *Rethinking the California Air Resources Board's Ozone Standards.*

139. J. F. Gent, E. W. Triche, T. R. Holford, et al., "Association of Low-Level Ozone and Fine Particles with Respiratory Symptoms in Children with Asthma," *Journal of the American Medical Association* 290 (2003): 1859–67.

140. U.S. Department of Health and Human Services, National Institute of Environmental Health Sciences, "NIEHS-Funded Researchers Find Low-Level Ozone Increases Respiratory Risk of Asthmatic Children," October 7, 2003, http://www.niehs.nih.gov/oc/news/yalasth.htm (accessed September 18, 2006).

141. These were the median rates reported in the study.

142. For data on asthma symptoms by month, see, for example, Gent, Triche, Holford, et al., "Association of Low-Level Ozone and Fine Particles"; Spokane Regional Health District, *Asthma in Spokane County*, June 2006, http://www.srhd.org/downloads/info_pubs/factsheets/Asthma2006FactSheet.pdf (accessed December 5, 2006); K. Tippy and N. Sonnenfeld, *Asthma Status Report, Maine 2002* (Augusta, Maine: Maine Bureau of Health, November 25, 2002); Michigan Department of Community Health, *An Analysis of Childhood Asthma Hospitalizations and Deaths in Michigan, 1989–1993*, by K. R. Wilcox and J. Hogan, summary and recommendations, http://www.michigan.gov/documents/Childhood_Asthma_6549_7.pdf (accessed September 18, 2006).

143. For data on asthma emergency room visits and hospitalizations by month, see, for example, Spokane Regional Health District, *Asthma in Spokane County*; California

Department of Health Services, *Environmental Health Investigations Branch, California County Asthma Hospitalization Chart Book: Data from 1998–2000*, by J. Stockman, N. Shaikh, J. Von Behren, et al., September 2003, http://www.ehib.org/cma/papers/Hosp_Cht_Book_2003.pdf (accessed September 18, 2006); Texas Department of Health, *Asthma Prevalence, Hospitalizations and Mortality—Texas, 1999–2001*, November 21, 2003, http://www.texasasthma.org/pdf/asthmafactsheet.pdf (accessed December 5, 2006); Tippy and Sonnenfeld, *Asthma Status Report*, Maine 2002; Michigan Department of Community Health, *An Analysis of Childhood Asthma Hospitalizations*.

144. UCLA Center for Health Policy Research, "Asthma Symptom Prevalence in California in 2001," in *2001 Health Interview* Survey, Spring/Summer 2002, http://www.healthpolicy.ucla.edu/pubs/files/Asthma-by-county-052002.pdf (accessed September 18, 2006).

145. U.S. Department of Health and Human Services, National Institute of Environmental Health Sciences, "In Young Rhesus Monkeys Smog Shown to Set up Lungs for Asthma," press release, October 12, 2000, http://www.niehs.nih.gov/oc/news/davisozo.htm (accessed September 18, 2006).

146. N. W. Johnston, S. L. Johnston, J. M. Duncan, et al., "The September Epidemic of Asthma Exacerbations in Children: A Search for Etiology," *Journal of Allergy and Clinical Immunology* 115 (2005): 132–38.

147. Schwartz, *Rethinking the California Air Resources Board's Ozone Standards*; California Air Resources Board, *Review of the California Ambient Air Quality Standard for Ozone*.

148. According to the CARB report cited in the previous note, going from recent ozone levels to statewide attainment of its new 8-hour standard would eliminate 3.7 million absence days per year. CARB also estimates that students are absent an average of six schooldays per year. There are about 6.9 million students in California's primary and secondary schools (including public and private), and the average school year is about 180 days. Based on these values, CARB implicitly estimates that attaining its standard will reduce school absences by nearly 9 percent.

149. F. D. Gilliland, K. Berhane, E. B. Rappaport, et al., "The Effects of Ambient Air Pollution on School Absenteeism Due to Respiratory Illness," *Epidemiology* 12 (2001): 43–54.

150. If, on the other hand, ozone reductions are equally effective in reducing the risk of absence due to any illness, then attaining CARB's standard would reduce illness-related absences by 16 percent.

151. California Air Resources Board, *Review of the California Ambient Air Quality Standard for Ozone*; Schwartz, *Rethinking the California Air Resources Board's Ozone Standards*.

152. Gilliland, Berhane, Rappaport, et al., "The Effects of Ambient Air Pollution on School Absenteeism."

153. K. Berhane and D. C. Thomas, "A Two-State Model for Multiple Time Series Data of Counts," *Biostatistics* 3 (2002): 21–32; V. Rondeau, K. Berhane, and D. C.

Thomas, "A Three-Level Model for Binary Time-Series Data: The Effects of Air Pollution on School Absences in the Southern California Children's Health Study," *Statistics in Medicine* 24 (2005): 1103–15.

154. H. Gong Jr., C. Sioutas, and W. S. Linn, "Controlled Exposures of Healthy and Asthmatic Volunteers to Concentrated Ambient Particles in Metropolitan Los Angeles," *Research Report/Health Effects Institute* 118 (2003): 1–36; discussion 37–47.

155. For example, between February 2005 and January 2006, Riverside, California, had a total of one hour with $PM_{2.5}$ greater than 200 µg/m³, and six hours with $PM_{2.5}$ greater than 150 µg/m³. Bakersfield, California had three hours with $PM_{2.5}$ greater than 150 µg/m³ and 200 µg/m³.

156. S. T. Holgate, T. Sandstrom, A. J. Frew, et al., "Health Effects of Acute Exposure to Air Pollution. Part I: Healthy and Asthmatic Subjects Exposed to Diesel Exhaust," *Research Report/Health Effects Institute* 112 (2003): 1–30; discussion 51–67.

157. Health Effects Institute, *Synopsis of Research Report 112: Effects of Particles on Lung Inflammation in Healthy and Asthmatic Volunteers*, January 2004, http://pubs.healtheffects.org/getfile.php?u=161, 2.

158. California Air Resources Board, *Review of the California Ambient Air Quality Standard for Ozone*, 2–9.

159. D. H. Horstman, L. J. Folinsbee, P. J. Ives, et al., "Ozone Concentration and Pulmonary Response Relationships for 6.6-Hour Exposures with Five Hours of Moderate Exercise to 0.08, 0.10, and 0.12 ppm," *American Review of Respiratory Disease* 142 (1990): 1158–63; S. M. Horvath, J. F. Bedi, D. K. Drechsler-Parks, and R. E. Williams, "Alterations in Pulmonary Function Parameters During Exposure to 80 ppb Ozone for 6.6 Hours in Healthy Middle Age Individuals" (paper, *Tropospheric Ozone and Environment: Papers from an International Conference,* Air and Waste Management Association, Los Angeles, March 1991, 59–70); W. F. McDonnell, H. R. Kehrl, S. Abdul-Salaam, et al., "Respiratory Response of Humans Exposed to Low Levels of Ozone for 6.6 Hours," *Archives of Environmental Health* 46 (1991): 145–50.

160. See, for example, L. J. Liu, R. Delfino, and P. Koutrakis, "Ozone Exposure Assessment in a Southern California Community," *Environmental Health Perspectives* 105 (1997): 58–65; R. J. Delfino, B. D. Coate, R. S. Zeiger, et al., "Daily Asthma Severity in Relation to Personal Ozone Exposure and Outdoor Fungal Spores," *American Journal of Respiratory and Critical Care Medicine* 154 (1996): 633–41; A. S. Geyh, J. Xue, H. Ozkaynak, and J. D. Spengler, "The Harvard Southern California Chronic Ozone Exposure Study: Assessing Ozone Exposure of Grade-School-Age Children in Two Southern California Communities," *Environmental Health Perspectives* 108 (2000): 265–70; T. Johnson, K. Clark, K. Anderson, et al., "A Pilot Study of Los Angeles Personal Ozone Exposures During Scripted Activities" (paper, Measurement of Toxic and Related Air Pollutants, Air and Waste Management Association, Research Triangle Park, N.C., May 7–9, 1996); K. Lee, W. J. Parkhurst, J. Xue, et al., "Outdoor/Indoor/Personal Ozone Exposures of Children in Nashville, Tennessee," *Journal of the Air and Waste Management Association* 54 (2004): 352–59;

M. S. O'Neill, M. Ramirez-Aguilar, F. Meneses-Gonzalez, et al., "Ozone Exposure Among Mexico City Outdoor Workers," *Journal of the Air and Waste Management Association* 53 (2003): 339–46.

161. A. R. Leston, W. M. Ollison, C. W. Spicer, and J. Satola, "Potential Interference Bias in Ozone Standard Compliance Monitoring," *Journal of the Air and Waste Management Association* 55 (2005): 1464–72.

162. California Air Resources Board, *Review of the California Ambient Air Quality Standard for Ozone.*

163. CARB's assumed background of 0.04 ppm is the average background level. But compliance with the standard depends on peak daily ozone levels in a given year. Therefore, the relevant question is what is the highest daily background level. This "peak" background level could well be much greater than 0.04 ppm.

164. W. C. Adams, "Comparison of Chamber and Face-Mask 6.6 Hour Exposures to Ozone on Pulmonary Function and Symptom Responses," *Inhalation Toxicology* 14 (2002): 745–64.

165. South Coast Air Quality Management District, *Multiple Air Toxics Exposure Study in the South Coast Air Basin* (Diamond Bar, Calif.: South Coast Air Quality Management District, March 2000).

166. National Cancer Institute, *SEER Cancer Statistics Review 1975–2001*, 2004, Table I–15, http://seer.cancer.gov/csr/1975_2001/results_merged/topic_lifetime_risk.pdf (accessed September 18, 2006).

167. For example, Doll and Peto estimated in 1981 that environmental pollution of all kinds accounted for 2 percent of all cancers. EPA estimated in 1987 that environmental chemicals contributed 1–3 percent of all cancers. Based on EPA's risk estimates for environmental pollutants, Michael Gough calculated that EPA could eliminate at most 0.25–1.3 percent of all cancers if its programs were 100 percent effective in reducing carcinogenic chemicals in the environment. These estimates include cancers from all types of environmental exposures to manmade chemicals, so air pollution would account for only a portion of the total amount of environmental cancer. R. Doll and R. Peto, "The Causes of Cancer: Quantitative Estimates of Avoidable Risks of Cancer in the United States Today," *Journal of the National Cancer Institute* 66 (1981): 1191–1308; U.S. Environmental Protection Agency, *Unfinished Business: A Comparative Assessment of Environmental Problems* (Washington, D.C.: Government Printing Office, February 1987); M. Gough, "How Much Cancer Can EPA Regulate Away?" *Risk Analysis* 10 (1990): 1–6.

168. If the lifetime risk of cancer from air pollution is 1/600, then in a cohort of 300 million people, roughly the current U.S. population, we'd expect 500,000 cancer cases over a seventy-five-year period, which is roughly the average lifespan of people in the cohort. Thus we would expect 500,000/75 or 6,666 new cancer cases per year, on average.

169. B. N. Ames and L. S. Gold, "The Causes and Prevention of Cancer: Gaining Perspective," *Environmental Health Perspectives* 105, supp. 4 (1997): 865–73; B. N.

Ames and L. S. Gold, "The Causes and Prevention of Cancer: The Role of Environment," *Biotherapy* 11 (1998): 205–20; B. N. Ames and L. S. Gold, "Paracelsus to Parascience: The Environmental Cancer Distraction," *Mutation Research* 447 (2000): 3–13.

170. Ames and Gold, "The Causes and Prevention of Cancer: Gaining Perspective."

171. Ames and Gold, "The Causes and Prevention of Cancer: The Role of Environment."

172. Ibid.

173. United Nations Environment Programme, *Scientific Assessment of Ozone Depletion: 2002*, 2002, http://www.wmo.ch/web/arep/reports/o3_assess_rep_2002_front_page.html (accessed November 27, 2006).

174. U.S. Environmental Protection Agency, "Improved Estimate of Non-Melanoma Skin Cancer Increases Increases [sic] Associated with Proposed Tropospheric Ozone Reductions," draft, May 22, 1997, http://www.aei-brookings.org/admin/authorpdfs/page.php?id=59 (accessed September 18, 2006). It suggests that average summer ozone levels would need to be reduced by between one and a few ppb in most 8-hour nonattainment areas in order to attain the standard. Also see R. Lutter and H. Gruenspect, "Assessing Benefits of Ground Level Ozone: What Role for Science in Setting National Ambient Air Quality Standards?" *Tulane Environmental Law Journal* 15 (2001): 85–96.

175. U.S. Department of Energy, *EPA Docket a-95-54, Iv-D-2694, Appendix B-9* (Washington, D.C.: Government Printing Office, March 21, 1995), cited in Lutter and Gruenspect, "Assessing Benefits of Ground Level Ozone." The DOE estimates were actually higher than this, but DOE assumed a 0.01 ppm reduction in seasonal average ozone levels. Smaller reductions would be necessary to attain the 8-hour standard in most areas, and we have adjusted the estimates downward to reflect this.

176. U.S. Environmental Protection Agency, *Air Quality Criteria for Ozone*, 10–35.

177. EPA never made this analysis public (see U.S. Environmental Protection Agency, "Improved Estimate of Non-Melanoma Skin Cancer"), but it was anonymously placed in a U.S. Office of Management and Budget docket and is now available online.

178. See R. Lutter, "Head in the Clouds Decisionmaking: EPA's Air Quality Standards for Ozone," in *Painting the White House Green: Rationalizing Environmental Policy Inside the Executive Office of the President*, eds. R. Lutter and J. F. Shogren (Washington, D.C.: Resources for the Future, 2004); U.S. Department of Energy, *EPA Docket A-95-54, IV-D-2694, Appendix B-9*; Lutter and Gruenspect, "Assessing Benefits of Ground Level Ozone." The United Nations Environment Programme (UNEP) also had no trouble with the idea that decreasing ozone increases harm from UV exposure. UNEP estimated that each 1 percent decrease in total atmospheric ozone results in a 1–2 percent increase in human exposure to UV light, and thereby to an increased incidence of cancer. According to UNEP, each 1 percent increase in UV exposure would likely result in ten to twenty new skin cancers per year per million people, or 3,000 to 6,000 new cancers in a population the size of the United States; United Nations Environment Programme, *Environmental Effects of Ozone Depletion: 1998*

Assessment, 1998, http://www.gcrio.org/UNEP1998/UNEP98.html (accessed September 18, 2006). The chapters of this report were also published as separate papers in the October 1998 issue of *Journal of Photochemistry and Photobiology B*, available at http://www.gcrio.org/ozone/toc.html (accessed November 27, 2006).

179. California Air Resources Board, *Review of the California Ambient Air Quality Standard for Ozone*, Presentation to the Governing Board, April 28, 2005, http://www.arb.ca.gov/research/aaqs/ozone-rs/4-28-05pres.pdf (accessed September 18, 2006).

180. It should also be noted that regulatory cancer-risk estimates include assumptions that guarantee a substantial overestimate of the true risk. Thus, CARB regulates cancer risks that are in reality orders of magnitude lower than one in one million. See, for example, Ames and Gold, "The Causes and Prevention of Cancer: Gaining Perspective."

181. Schwartz, *Rethinking the California Air Resources Board's Ozone Standards*.

182. U.S. Environmental Protection Agency, *Mercury Study Report to Congress, Volume I: Executive Summary*, December 1997, http://www.epa.gov/ttn/oarpg/t3/reports/volume1.pdf (accessed September 18, 2006); K. R. Mahaffey, R. P. Clickner, and C. C. Bodurow, "Blood Organic Mercury and Dietary Mercury Intake: National Health and Nutrition Examination Survey, 1999 and 2000," *Environmental Health Perspectives* 112 (2004): 562–70.

183. C. Seigneur, K. Vijayaraghavan, K. Lohman, et al., "Global Source Attribution for Mercury Deposition in the United States," *Environmental Science & Technology* 38 (2003): 555–69.

184. G. J. Myers and P. W. Davidson, "Does Methylmercury Have a Role in Causing Developmental Disabilities in Children?" *Environmental Health Perspective* 108, supp. 3 (2000): 413–20.

185. U.S. Department of Health and Human Services, National Institutes of Health, "Toxicology Glossary," http://sis.nlm.nih.gov/enviro/glossaryr.html (accessed November 27, 2006).

186. J. Lowy, "EPA Raises Estimate of Newborns Exposed to Mercury," Scripps-Howard News Service, April 4, 2004.

187. Friends of the Earth, "Hard-Hitting Ad Tells President Bush to Protect America's Children from Mercury Pollution," press release, March 16, 2004, http://www.foe.org/new/releases/304mercpr.html (accessed August 30, 2004); Natural Resources Defense Council, "Mercury Contamination in Fish: A Guide to Staying Healthy and Fighting Back," http://www.nrdc.org/health/effects/mercury/index.asp (accessed September 18, 2006); J. Morris, "EPA to Investigate Its Proposed Mercury Rule," *Dallas Morning News*, May 13, 2004; J. Nesmith, "Senators Attack Mercury Proposal; EPA Accused of Pro-Industry Bias," *Atlanta Journal-Constitution*, April 13, 2004; Lowy, "EPA Raises Estimate of Newborns Exposed to Mercury."

188. Centers for Disease Control and Prevention, National Center for Health Statistics. *National Health and Nutrition Examination Survey Data*, http://www.cdc.gov/nchs/nhanes.htm (accessed August 16, 2006).

189. U.S. Environmental Protection Agency, "Methylmercury (MeHg) (CASRN 22967-92-6)," Integrated Risk Information System, July 27, 2001, http://www.epa.gov/iris/subst/0073.htm (accessed September 18, 2006).

190. Strictly speaking, 85 ppb is called the Benchmark Dose, or BMD. The bottom of the 95 percent confidence interval is called the Benchmark Dose Lower Limit, or BMDL. EPA then divides the BMDL by safety factors to arrive at the Reference Dose, or RfD.

191. U.S. Environmental Protection Agency, "Methylmercury (Mehg) (CASRN 22967-92-6)."

192. Centers for Disease Control, "Blood Mercury Levels in Young Children and Childbearing-Aged Women—United States, 1999–2002," *Morbidity and Mortality Weekly Report* 53 (2004): 1018–20; Centers for Disease Control, National Center for Health Statistics, "National Health and Nutrition Examination Survey," http://www.cdc.gov/nchs/nhanes.htm (accessed November 27, 2006).

193. Centers for Disease Control, National Center for Health Statistics, *National Health and Nutrition Examination Survey Data.*

194. Ibid.

195. P. Grandjean, E. Budtz-Jorgensen, R. F. White, et al., "Methylmercury Exposure Biomarkers as Indicators of Neurotoxicity in Children Aged 7 Years," *American Journal of Epidemiology* 150 (1999): 301-5; P. Grandjean, P. Weihe, R. F. White, et al., "Cognitive Deficit in 7-Year-Old Children with Prenatal Exposure to Methylmercury," *Neurotoxicology and Teratology* 19 (1997): 417–28.

196. G. J. Myers, P. W. Davidson, C. Cox, et al., "Prenatal Methylmercury Exposure from Ocean Fish Consumption in the Seychelles Child Development Study," *Lancet* 361 (2003): 1686–92.

197. U.S. Environmental Protection Agency, "Methylmercury (MeHg)."

198. Myers, Davidson, Cox, et al., "Prenatal Methylmercury Exposure."

199. This was true for the full sample of children. However, there were statistically significant decreases in scores on a few of the administered tests when one child was removed from the dataset. This child had the highest mercury exposure, but normal test scores. However, even with this child removed, there was no decrease in scores on any portion of the IQ test. K. S. Crump, T. Kjellstrom, A. M. Shipp, et al., "Influence of Prenatal Mercury Exposure Upon Scholastic and Psychological Test Performance: Benchmark Analysis of a New Zealand Cohort," *Risk Analysis* 18 (1998): 701–13.

200. The shark eaten by New Zealanders has about seven times the mercury concentration as the fish eaten by the Seychellois; Myers, Davidson, Cox, et al., "Prenatal Methylmercury."

201. D. C. Rice, R. Schoeny, and K. Mahaffey, "Methods and Rationale for Derivation of a Reference Dose for Methylmercury by the U.S. EPA," *Risk Analysis* 23 (2003): 107–15.

202. Grandjean, Weihe, White, et al., "Cognitive Deficit in 7-Year-Old Children with Prenatal Exposure to Methylmercury."

203. Nesmith, "Senators Attack Mercury Proposal."

204. Morris, "EPA to Investigate Its Proposed Mercury Rule."

205. S. Hartsoe, "Experts: EPA Emissions Plan Harmful to Health, NC Tourism," Associated Press, February 26, 2004.

206. Natural Resources Defense Council, "Mercury Contamination in Fish."

207. Friends of the Earth, "Hard-Hitting Ad Tells President Bush to Protect America's Children."

208. Elizabeth Weise, "Eating Fish: There's A Catch," *USA Today*, October 26, 2005, 1D.

209. S. E. Dudley, *National Ambient Air Quality Standard for Ozone*, Mercatus Center, George Mason University, March 12, 1997, http://www.mercatus.org/repository/docLib/20060830_MR_RSP_PIC_EPA_NAAQS_Dudley_March_12_1997.pdf (accessed December 5, 2006); S. Huebner and K. Chilton, *EPA's Case for New Ozone and Particulate Standards: Would Americans Get Their Money's Worth?* Center for the Study of American Business, Washington University in St. Louis, June 1997, csab.wustl.edu/csab/CSAB%20pubs-pdf%20files/Policy%20Studies/PS139%20Huebner-Chilton.pdf (accessed September 18, 2006); Lutter, *Is EPA's Ozone Standard Feasible?*; Lutter, "Head in the Clouds Decision-Making."

210. Lutter, *Is EPA's Ozone Standard Feasible?*; Winner and Cass, "Effect of Emissions Control."

211. A. P. Bartel and L. G. Thomas, "Predation through Regulation: The Wage and Profit Effects of the Occupational Safety and Health Administration and the Environmental Protection Agency," *Journal of Law and Economics* 30 (1987): 239-64; D. Schoenbrod, "Protecting the Environment in the Spirit of the Common Law," in *The Common Law and the Environment: Rethinking the Statutory Basis for Modern Environmental Law*, ed. R. E. Meiners and A. P. Morriss (Lanham, Md.: Rowman & Littlefield, 2000); A. Wildavsky, *Searching for Safety* (New Brunswick, N.J.: Transaction Publishers, 1988). The costs of environmental regulations are also regressive, falling most heavily on the poorest. See F. B. Cross, "When Environmental Regulations Kill: The Role of Health/Health Analysis," *Ecology Law Quarterly* 22 (1995): 729–84; H. D. Robinson, "Who Pays for Industrial Pollution Abatement?" *Review of Economics and Statistics* 67 (1985): 702–6.

212. R. Lutter, J. Morrall III, and W. Viscusi, "The Cost-Per-Life-Saved Cutoff for Safety-Enhancing Regulations," *Economic Inquiry* 37 (1999): 599–608; W. K. Viscusi, "The Value of Risks to Life and Health," *Journal of Economic Literature* 31 (1993): 1912–46; Wildavsky, *Searching for Safety*.

213. Lutter, Morrall, and Viscusi, "The Cost-Per-Life-Saved Cutoff." The value is adjusted from 1997 to 2004 dollars based on the CPI.

214. Based on estimates cited in note 211.

215. T. O. Tengs, M. E. Adams, J. S. Pliskin, et al., "Five-Hundred Life-Saving Interventions and Their Cost-Effectiveness," *Risk Analysis* 15 (1995): 369–90.

Chapter 8: Has the Clean Air Act Been Good for Americans?

1. D. Schoenbrod, "Putting the 'Law' Back into Environmental Law," *Regulation* (Spring 1999), http://www.cato.org/pubs/regulation/regv22n1/envirolaw.pdf.

2. National Research Council, *Air Quality Management in the United States*.

3. A. P. Morriss, "The Politics of the Clean Air Act," in *Political Environmentalism: Going Behind the Green Curtain*, ed. T. L. Anderson (Stanford, Calif.: Hoover, 2000).

4. In a survey of state air agencies, Morriss (ibid.) found that more than eight hundred full-time-equivalent personnel are devoted to maintaining states' SIPs.

5. National Research Council, *Modeling Mobile-Source Emissions* (Washington, D.C.: National Academies Press, 2000); A. Russell and R. Dennis, "NARSTO Critical Review of Photochemical Models and Modeling," *Atmospheric Environment* 34 (2000): 2283–2324; R. F. Sawyer, R. A. Harley, S. H. Cadle, et al., "Mobile Sources Critical Review: 1998 NARSTO Assessment," *Atmospheric Environment* 34 (2000): 2161–81.

6. Watson, Chow, and Fujita, "Review of Volatile Organic Compound Source Apportionment."

7. D. R. Lawson, P. A. Walsh, and P. Switzer, *Analysis of U.S. Roadside Vehicle Emissions and Tampering Survey Data and Evaluation of Inspection and Maintenance Programs*, Final Report (Alpharetta, Ga.: Coordinating Research Council, March 21, 1996); National Research Council, *Evaluating Vehicle Emissions Inspection and Maintenance Programs* (Washington, D.C.: National Academy Press, 2001); J. Schwartz, *An Analysis of the USEPA's 50-Percent Discount for Decentralized Vehicle I/M Programs*, California Inspection and Maintenance Review Committee, March 1995, http://www.joelschwartz.com/pdfs/50pct_Schwartz95.pdf (accessed November 27, 2006); J. Schwartz, *Improving Evaluation of Mobile Source Policies: Comments to the National Research Council on Its Review of EPA's Mobile Source Emissions Factor Model*, California Inspection and Maintenance Review Committee, March 1999, http://www.joelschwartz.com/pdfs/NRC_MOBILE_Schwartz99.pdf (accessed November 27, 2006) ; D. H. Stedman, G. A. Bishop, P. Aldrete, and R. S. Slott, "On-Road Evaluation of an Automobile Emission Test Program," *Environmental Science & Technology* 31 (1997): 927–31; D. H. Stedman, G. A. Bishop, and R. S. Slott, "Repair Avoidance and Evaluating Inspection and Maintenance Programs," *Environmental Science & Technology* 32 (1998): 1544–45.

8. D. R. Lawson, P. J. Groblicki, D. H. Stedman, et al., "Emissions from In-Use Motor Vehicles in Los Angeles: A Pilot Study of Remote Sensing and the Inspection and Maintenance Program," *Journal of the Air & Waste Management Association* 40 (1990): 1096–1105; National Research Council, *Evaluating Vehicle Emissions Inspection and Maintenance Programs*; S. P. Beaton, G. A. Bishop, Y. Zhang, et al., "On-Road Vehicle Emissions: Regulations, Costs, and Benefits," *Science* 268 (1995): 991–92; Calvert, Heywood, Sawyer, et al., "Achieving Acceptable Air Quality"; A. Pollack, "Comparison of Remote Sensing Data and Emission Factor Model Predictions" (paper, Second Annual CRC On-Road Emissions Workshop, Coordinating Research Council, San Diego, December 1–3, 1992); D. H. Stedman, G. Bishop, J. E. Peterson, and P. L.

Guenther, *On-Road CO Remote Sensing in the Los Angeles Basin—Final Report* (Sacramento: California Air Resources Board, August 1991); D. H. Stedman, G. A. Bishop, J. E. Peterson, et al., *On-Road Carbon Monoxide and Hydrocarbon Remote Sensing in the Chicago Area* (Chicago: Illinois Department of Energy and Natural Resources, October 1991); Schwartz, *No Way Back*; D. H. Stedman, *Remote Sensing: A New Tool for Automobile Inspection & Maintenance*, Independence Institute, January 2002, http://www.feat.biochem.du.edu/assets/reports/Ind_Inst_IM_position_1_2002.pdf (accessed September 20, 2006); D. H. Stedman, G. A. Bishop, S. P. Beaton, et al., *On-Road Remote Sensing of CO and HC Emissions in California—Final Report* (Sacramento: California Air Resources Board, February 1994); California Air Resources Board, *Evaluation of ARB's In-Use Vehicle Surveillance Program.*

9. L. G. Wayne and Y. Horie, *Evaluation of ARB's In-Use Vehicle Surveillance Program*, Final Report (Sacramento: California Air Resources Board, October 1983).

10. Different groups of cars constitute the worst 5 percent for each pollutant, but there is substantial overlap; Schwartz, *No Way Back.*

11. Schwartz, *An Analysis of the USEPA's 50-Percent Discount*; Stedman, Bishop, Aldrete, et al., "On-Road Evaluation of an Automobile Emission Test Program"; Stedman, Bishop, and Slott, "Repair Avoidance"; T. P. Wenzel, "Evaluating the Long-Term Effectiveness of the Phoenix IM240 Program," *Environmental Science and Policy* 4 (2001): 377–89; T. P. Wenzel, "Use of Remote Sensing Measurements to Evaluate Vehicle Emission Monitoring Programs: Results from Phoenix, Arizona," *Environmental Science and Policy* 6 (2003): 153–66.

12. D. R. Lawson, S. Diaz, E. M. Fujita, et al., *Program for the Use of Remote Sensing Devices to Detect High-Emitting Vehicles, Prepared for the South Coast Air Quality Management District* (Reno: Desert Research Institute, April 16, 1996); National Research Council, *Evaluating Vehicle Emissions Inspection and Maintenance Programs*; Schwartz, *No Way Back*; Stedman, *Remote Sensing*; D. H. Stedman, G. A. Bishop, J. E. Peterson, and T. Hosick, *Provo Pollution Prevention Program: A Pilot Study of the Cost-Effectiveness of an On-Road Vehicle Emissions Reduction Program* (Provo, Utah: University of Denver, January 15, 1993).

13. H. K. Gruenspect and R. N. Stavins, "New Source Review under the Clean Air Act: Ripe for Reform," *Resources* (Spring 2002): 19–23; B. Swift, "How Environmental Laws Work: An Analysis of the Utility Sector's Response to Regulation of Nitrogen Oxides and Sulfur Dioxide under the Clean Air Act," *Tulane Environmental Law Journal* 14 (2001): 309–425.

14. B. Swift, "Grandfathering, New Source Review, and NOx—Making Sense of a Flawed System," *Environment Reporter* 31 (2000): 1588–96.

15. Ibid.

16. Ibid.

17. R. F. Kennedy Jr., "Profits, Politics, and Pollution," speech given at the Michigan Theater, Ann Arbor, Mich., November 10, 2005, http://www.ecocenter.org/200601/rfk-jr_speech_200601.shtml.

18. Schoenbrod, "Protecting the Environment in the Spirit of the Common Law."

19. Morriss, "The Politics of the Clean Air Act."

20. J. Adler, "Rent Seeking Behind the Green Curtain," *Regulation* 19, no. 4 (1996), http://www.cato.org/pubs/regulation/reg19n4b.html (accessed November 27, 2006).

21. J. Adler, "Clean Politics, Dirty Profits: Rent-Seeking Behind the Green Curtain," in *Political Environmentalism: Going Behind the Green Curtain* (Stanford: Hoover Institution Press, 2000).

22. B. Egelko, "Court Gives State Hope for Waiver on Ethanol; EPA Ordered to Reconsider Refusal of Request," *San Francisco Chronicle*, July 18, 2003, A25; Richard Barrett and Donald Stedman, "High Temperature and Fuel Impact on HC Emissions" (paper, Fourteenth CRC On-Road Vehicle Emissions Workshop, San Diego, Calif., March 29–31, 2004), www.feat.biochem.du.edu/assets/posters/Barrett_Stedman _CRC04_HC_poster.pdf (accessed November 27, 2006).

23. *Energy Policy Act of 2005*, Public Law 109-58, 109th Congress, http://frwebgate.access.gpo.gov/cgi-bin/getdoc.cgi?dbname=109_cong_public_laws&docid=f:publ058.109 (accessed November 27, 2006).

24. Ibid. For context, American motorists use a total of about 135 billion gallons of gasoline each year, so the 2012 ethanol mandate represents about a little over 5 percent of current consumption by volume. But note that ethanol has only about two-thirds the energy content of gasoline per gallon, so it takes about 1.5 gallons of ethanol for a vehicle to be driven the same distance as it can be driven on one gallon of gasoline.

25. The subsidy was first enacted in 1979 and has ranged from forty to sixty cents per gallon of ethanol. U.S. General Accounting Office, "Petroleum and Ethanol Fuels: Tax Incentives and Related GAO Work," September 25, 2000, http://www.gao.gov/new.items/rc00301r.pdf (accessed November 27, 2006).

26. T. J. Zywicki, "Industry and Environmental Lobbyists: Enemies or Allies?" in *The Common Law and the Environment.*

27. Adler, "Clean Politics, Dirty Profits."

28. The story is told in detail in B. A. Ackerman and W. T. Hassler, *Clean Coal/Dirty Air, or How the Clean Air Act Became a Multi-Billion-Dollar Bail-Out for High-Sulfur Coal Producers and What Should Be Done About It* (New Haven, Conn.: Yale University Press, 1981).

29. Adler, "Clean Politics, Dirty Profits."

30. Zywicki, "Industry and Environmental Lobbyists."

31. Ibid.

32. Ibid.

33. B. L. Benson, *Unnatural Bounty: Distorting the Incentives of Major Environmental Groups* (Bozeman, Mont.: Property and Environment Research Center, July 2006), http://www.perc.org/pdf/ps37.pdf (accessed November 27, 2006).

34. Ibid.

35. Ibid.

36. Adler, "Clean Politics, Dirty Profits."

37. J. Adler, "Putting EPA to the I/M Grant Test," *Washington Times*, December 12, 1995, A16; T. Frank, "Car-Test Firm Has Big Clout; Envirotest Founder Uses Government," *Denver Post*, July 28, 1996, A-01; U.S. General Accounting Office, *EPA's Grants and Agreements* (Washington, D.C.: Government Printing Office, November 16, 1995); D. Hopey, "Donors Annoyed About Support; Local Lung Association Cancels Newspaper Ad for E-Check Auto Test," *Pittsburgh Post-Gazette*, September 28, 1994, C1; S. Voas, "Clean Air Ad Series Riles Some Legislators," *Pittsburgh Post-Gazette*, November 16, 1994, B8.

38. Schoenbrod, "Protecting the Environment."

39. Zywicki, "Industry and Environmental Lobbyists."

40. However, it should be noted that cap-and-trade programs themselves create opportunities for political manipulation through the system for allocating pollution permits when the program is launched. As a result, many observers prefer pollution taxes to cap-and-trade.

41. See, for example, U.S. Environmental Protection Agency, "EPA Funds New Research on Air Pollution, Children's Health and Watershed Protection," press release, July 29, 2003, http://yosemite.epa.gov/opa/admpress.nsf/4a4be1d8ed4c0b 198525702100564fe3/0c118e5e18c3d34a85256d720051db49!OpenDocument (accessed November 28, 2006); U.S. Environmental Protection Agency, "Renewed EPA Funding for Harvard School of Public Health Research Focuses on Major Component of Air Pollution," press release, December 15, 2005, http://yosemite. epa.gov/opa/admpress.nsf/d9bf8d9315e942578525701c005e573c/96125238fcf089 b2852570dc0051746c!OpenDocument (accessed November 28, 2006); U.S. Environmental Protection Agency, "EPA Awards $8 Million to University of Rochester for Study of Air Pollution Health Effects," press release, December 12, 2005; University of California at Davis, "UC Davis Wins $8 Million EPA Grant to Study Health Effects of Air Pollution," press release, November 15, 2005, http://www.news. ucdavis.edu/search/news_detail.lasso?id=7541. Also see EPA's list of EPA-funded air pollution health research centers at http://cfpub.epa.gov/ncer_abstracts/index.cfm/ fuseaction/outlinks.centers (accessed November 28, 2006). The California Air Resources Board is also a major funder of air pollution health research. Details on CARB's research funding programs can be found at http://www.arb.ca.gov/research/ health/healthres.htm (accessed November 28, 2006).

42. U.S. Environmental Protection Agency, *Environmental Education Grants 1992–2005*, http://www.epa.gov/enviroed/grants/index.html#note (accessed November 28, 2006); Capital Research Center, *Database of Grantmakers and Non-Profits Receiving Grants*, http://www.greenwatch.org/search/search.asp (accessed November 28, 2006).

43. See, for example, Clean Air Task Force, *Children at Risk* (Boston: Clean Air Task Force, 2002); Natural Resources Defense Council, *Heat Advisory* (New York: Natural Resources Defense Council, 2004); American Lung Association, "American Lung Association Calls on EPA to Strengthen Particle Pollution Protections for All

Americans," March 8, 2006; American Lung Association, "100+ Scientists Endorse Stringent New PM Standards," December 5, 2005, http://www.cleanairstandards.org/article/articleview/404/1/41/.

44. For general examples, see S. Inskeep, "Commuting IV," *All Things Considered*, National Public Radio, May 30, 1997, http://www.npr.org/templates/story/story.php?storyId=1039432 (accessed November 27, 2006), and R. O'Toole, *The Vanishing Automobile* (Bandon, Ore.: Thoreau Institute, 2000), 260–61. Dom Nozzi of the organization Walkable Streets said, "It is a serious strategic blunder for sprawl-busters and other community and environmental advocates to oppose traffic congestion," D. Nozzi, "Traffic Congestion: Friend or Foe?" Walkable Streets, undated, http://www.walkablestreets.com/congest.htm (accessed November 27, 2006); Marie Howland, director of the Urban Studies and Planning Program at the University of Maryland, said, "My hope is that it will get so congested that people will want to invest in public transit," J. Hernandez, "Fewer Marylanders Riding Public Transportation to Work, Carpooling," Associated Press, May 30, 2002; Michael Replogle of the environmental group Environmental Defense argues, "Limiting further highway capacity expansion, reducing highway capacity, and calming traffic (especially in central areas) can be effective strategies for reducing energy use, air pollution, and other environmental problems," Transportation Research Board, "Minority Statement of Michael A. Replogle," in *Expanding Metropolitan Highways: Implications for Air Quality and Energy Use* (Washington, D.C.: National Academies Press, 1995), 358, http://www.nap.edu/books/0309061075/html (accessed November 27, 2006).

45. U.S. Environmental Protection Agency and Smart Growth Network, "Smart Growth Funding Resource Guide," June 2001, http://www.smartgrowth.org/pdf/funding_resources.pdf (accessed November 28, 2006); Smart Growth Network, "Technical Assistance Programs Offered by the Smart Growth Network," July 2006 (rev. ed.), http://www.smartgrowth.org/pdf/SGN_TA_JulyRevision2006.pdf (accessed November 28, 2006); R. O'Toole, "EPA Funds Anti-Sprawl Critics with Tax Dollars," Cato Institute, "Daily Commentary," February 25, 2000, http://www.cato.org/dailys/02-25-00.html (accessed November 28, 2006).

46. Lest readers think that Americans are unique in favoring automobiles and suburbs, note that Europeans use automobiles for 78 percent of travel. Most Europeans also live in suburbs, and nearly all new development everywhere in the Western world is suburban-style development. Bureau of Transportation Statistics, *National Transportation Statistics 2004* (Washington, D.C.: U.S. Department of Transportation, January 2005), http://www.bts.gov/publications/national_transportation_statistics/2004/index.html (accessed November 27, 2006); European Commission, *Panorama of Transport: Statistical Overview of Transport in the European Union*, Part 2, 2003, http://epp.eurostat.cec.eu.int/cache/ITY_OFFPUB/KS-DA-04-001-2/EN/KS-DA-04-001-2-EN.PDF (accessed November 27, 2006); W. Cox, "High-Income World Metropolitan Areas: Core City & Suburban Population Trends," *Demographia*, http://www.demographia.com/db-highmetro.htm (accessed November 27, 2006).

47. A. M. Howitt and A. Altschuler, "The Politics of Controlling Auto Air Pollution," in *Essays in Transportation Economics and Policy: A Handbook in Honor of John R. Meyer*, ed. J. Gomez-Ibanez, W. B. Tye, and C. Winston (Washington, D.C.: National Academies Press, 1999).

48. Lutter, *Is EPA's Ozone Standard Feasible?*; Winner and Cass, "Effect of Emissions Control"; Schwartz, *Rethinking the California Air Resources Board's Ozone Standard.*

49. S. Peltzman, "Regulation and the Natural Progress of Opulence," AEI-Brookings Joint Center for Regulatory Studies, 2004, http://www.aei-brookings.org/admin/authorpdfs/page.php?id=1144#search=%22peltzman%20natural%20progress%20of %20opulence%22 (accessed March 21, 2007).

50. Schoenbrod, "Protecting the Environment."

51. Ibid.

Appendix to Chapter 6

1. Actually, Manhattan had one ozone monitor that violated the standard. It was at the top of the World Trade Center and was destroyed on September 11, 2001. This monitor measured ozone levels more than 1,000 feet above ground level. Since ground level is where people breathe the air, we ignored these data. Incidentally, the American Lung Association gave Manhattan a failing grade for ozone in the first few editions of *State of the Air*, based on data from this high-altitude monitor.

2. These estimates were downloaded from U.S. Census Bureau, "County Population and Estimated Components of Population Change, All Counties: April 1, 2000 to July 1, 2005," 2006, http://www.census.gov/popest/counties/files/CO-EST 2005-ALLDATA.csv, (accessed November 27, 2006).

About the Authors

Joel M. Schwartz is a visiting fellow at the American Enterprise Institute, where he studies the science, policy, and politics of air pollution, climate change, and other environmental concerns. He is the author of the AEI study *No Way Back: Why Air Pollution Will Continue to Decline*, as well as dozens of other studies and articles on environmental policy.

Before coming to AEI, Mr. Schwartz directed the Reason Public Policy Institute's Air Quality Project. He also served as Executive Officer of the California Inspection and Maintenance Review Committee, a government agency charged with evaluating California's vehicle emissions inspection program and making recommendations to the legislature and governor on program improvements. He has also worked at the RAND Corporation, the South Coast Air Quality Management District, and the Coalition for Clean Air.

Mr. Schwartz holds a bachelor's degree in chemistry from Cornell University and a master's degree in planetary science from the California Institute of Technology. He was a German Marshall Fund fellow in 1993, during which he studied European approaches to transportation and air quality policy.

Steven F. Hayward, is the F. K. Weyerhaeuser Fellow in Law and Economics at the American Enterprise Institute in Washington, D.C. He holds a Ph.D in American Studies and an M.A. in Government from Claremont Graduate School. He is the author of four books on American and British politics, and writes frequently on a wide range of current topics, including environmentalism, law, economics, and public policy for publications including *National Review*, *The Weekly Standard*, *The Public Interest*, *The American Spectator*, *The Claremont Review of Books*, and *Policy Review*. His newspaper articles have appeared in the *New York Times*, the

Wall Street Journal, the *San Francisco Chronicle*, the *Chicago Tribune*, and dozens of other daily newspapers. He co-authors AEI's *Environmental Policy Outlook*, and is the principal author of the annual *Index of Leading Environmental Indicators*, published each year on Earth Day.

Index